WIRING & LIGHTING

WIRING & LIGHTING
JACKSON & DAY

HarperCollins*Publishers*

Published by HarperCollins Publishers
London

This book was created exclusively for
HarperCollins Publishers by
Jackson Day Jennings Ltd trading as
Inklink.

**Design, art direction
and project management**
Simon Jennings

Text
Albert Jackson
David Day

**Text and
editorial direction**
Albert Jackson

Illustrations editor
David Day

**Designer and
production assistant**
Alan Marshall

Illustrators
David Day
Robin Harris

Additional illustrations
Brian Craker
Michael Parr
Brian Sayers

First published in 1988
This edition published 1992,
reprinted in 1993, 1994

The text and illustrations in this book
were previously published in
Collins Complete DIY Manual

Copyright © 1986, 1988, 1992
HarperCollins Publishers

ISBN 0 00 412817 6

A catalogue record for this book is
available from the British Library

Printed and bound in Hong Kong

Picture credits
Crown Decorative Products Ltd: 46BR
Dimplex Heating Ltd: 59T
Michael Dunne/EWA: 45, 46TR
Michael Nicholson/EWA: 46BL
Neil Lorimer/EWA: 46TL
Paul Chave: 9, 12, 13, 14, 15, 17L, 18, 19, 20,
22R, 23, 27L, 27R, 28, 29, 34, 36, 37, 38, 40, 41,
44, 49, 54
Robin Harris: 17R, 22L, 27C
TI Creda Ltd: 59C, 59B

CONTENTS

Cross-references
There are few DIY projects that do not require a combination of skills. Decorating a single room, for instance, might also involve modifying the plumbing or electrical wiring, installing ventilation or insulation, repairing the structure of the building and so on. As a result, you might have to refer to more than one section of this book. To help you locate the relevant sections, a symbol (▷) in the text refers you to a list of cross-references in the page margin. Those references printed in bold type are directly related to the task in hand. Other references which will broaden your understanding of the subject are printed in light-weight type.

REDUCING THE COST OF ELECTRICITY

Pressures from all sides urge us to conserve energy – electricity as well as fossil fuels like coal, oil and gas – but even without such encouragement the totals on our quarterly electricity bills should be stimulus enough to make us find ways of using less power. No-one wants to live in a poorly heated, dismally lit house without the comforts of hot baths, TV, record players and other conveniences, but you may be able to identify where energy is wasted, then find ways to reduce waste without compromising your comfort or pleasure.

Avoid false economy

Whether you do your own wiring or employ a professional, don't try to economize by installing fewer sockets than you really need. When you rewire a room fit as many as you may possibly use. The inconvenience, later on, of running extra cable and disturbing decoration will far outweigh the cost of an extra socket or two.

Similarly, don't restrict your use of lighting unnecessarily. It uses relatively little power, so there is no point in risking, say, accidents on badly lit stairs. Nor need you strain your eyes in the glare from a single light hanging from the ceiling when extra lights can give comfortable and attractive background illumination where needed.

Fitting controls to save money

It will be clear from the chart (See opposite) that heating of one sort or another is the main consumer of power. One way to economize is to fit devices that are designed to regulate its use to suit your life style and keep your heating at comfortable but economic temperatures.

Thermostats
Most modern heating has some form of thermostatic control – a device that will switch power off when surroundings reach a certain temperature. Many thermostats are marked out simply to increase or decrease the temperature. In this case you have to experiment with various settings to find the one that suits you. If you can set it accurately try 18°C (65°F) for everyday use, though elderly people are more comfortable at about 21°C (70°F).

An immersion heater thermostat both saves money and prevents water becoming dangerously hot. Set it at 60°C (140°F). (For Economy 7 setting see right.)

Time switches
Even thermostatically controlled heating is expensive if run continuously but an automatic time switch can turn it on and off at pre-set times so that you get up in the morning or home in the evening to a warm house. Set it to turn off the heating a half-hour before you leave home or go to bed, as the house will take time to cool down.

A similar device will ensure water at its hottest when it is needed.

● **Insulation**
Measures taken to save energy are of little use unless you insulate the house as well as the hot water cylinder and pipework (◁). You can do most of the work yourself with very little effort or cost.

Recording consumption

Keep an accurate record of your energy saving by taking weekly readings. Note the dates of measures taken to cut consumption and compare the corresponding drop in meter readings.

Digital meters
Modern meters simply display a row of figures or digits that represent the total number of units consumed since the meter was installed. To deduce the number of units used since your last electricity bill subtract the 'present reading' shown on the bill from the number of units now shown on the meter. Make sure that the bill shows an actual reading and not an estimate, indicated by an E before the reading.

ECONOMICAL OFF-PEAK RATES

Electricity is normally sold at a general-purpose rate, every unit used costing the same, but if you warm your home with storage heaters and heat your water electrically you should consider the advantages of the economical off-peak rate. The system, called Economy 7, allows you to charge storage heaters and heat water at less than half the general-purpose rate for seven hours starting between midnight and 1 a.m. Any other appliance used in that time gets cheap power too, so more savings can be made by having a timer to turn on a dishwasher or automatic washing machine when the household has gone to bed. Economy 7 daytime rate is higher than the general-purpose one, but the cost of running 24-hour appliances like freezers and refrigerators is balanced because they too use cheap power for seven hours.

For full benefit from off-peak water heating use a 182 to 227 litre (40 to 50 gallon) cylinder to store as much cheap hot water as possible. You will need a twin-element heater or two separate units. One heater, near the base of the cylinder, heats the whole tank on cheap power; another, about half way, tops up the hot water during the day. Set the night-time heater about 5°C (10°F) higher than the upper heater.

The Electricity Board supplies those using the Economy 7 programme with a special meter to record daytime and night-time consumption separately, also a timer that switches the supply from one rate to the other.

HOW TO READ DIAL METERS

The principle of a dial meter is simple. Ignore the dial marked 1/10, which is only for testing. Start with the dial indicating single units (kWh) and, working from right to left, record the readings from 10, 100, 1000 and finally 10,000 units. Note the digits the pointers have passed. If a pointer is, say, between 5 and 6 record 5. If a pointer is right on a number, say 8, check the next dial on the right. If that point is between 9 and 0 record 7; if it is past 0 record 8.

Remember that adjacent dials revolve in opposite directions, alternating along the row.

Reading a dial meter
Write down your reading in reverse order – from right to left. This meter records 76,579 units.

RUNNING COSTS OF YOUR APPLIANCES

Apart from the standing charge or hire-purchase payments, your electricity bill is calculated from the number of units of electricity you have used in a given period. Each unit represents the amount used in one hour by a 1kW appliance. An appliance rated at 3kW uses the same amount of energy in 20 minutes.

TYPICAL RUNNING COSTS

Appliance	Typical usage	No. of units	Appliance	Typical usage	No. of units
Cooker	Cooks one day's meals for four people	2½	Iron	In use for 2 hours	1
Microwave	Cooks two joints of meat	1	Vacuum cleaner	Works for 1½-2 hours	1
Slow cooker	Cooks for 8 hours	1	Cooker hood	Runs for 24 hours continuously	2
Storage heater (2 kW)	Provides one day's heating	8½	Extractor fan	Runs for 24 hours continuously	1
Bar fire or fan heater (3 kW)	Gives heat for one hour	3	Hair dryer	Runs for 2 hours	1
Immersion heater	Supplies one day's hot water for family of four	9	Shaver	Provides 1800 shaves	1
Instant water heater	Heats 2 to 3 bowls of washing up water	1	Single overblanket	Warms the bed for one week	2
Instant shower	Delivers 1 to 2 showers	1	Single underblanket	Warms the bed for one week	1
Dishwasher	Washes one full load	2	Power drill	Works for 4 hours	1
Automatic washing machine	Deals with one full load with pre-wash	2½	Hedge trimmer	Trims for 2½ hours	1
Tumble dryer	Dries same load	2½	Cylinder lawn mower	Cuts grass for 3 hours	1
4 cu ft refrigerator	Keeps food fresh for one week	7	Hover mower	Cuts grass for 1 hour	1
6 cu ft freezer	Maintains required temperature for one week	9	Stereo system	Plays for 8 hours	1
Heated towel rail	Warms continuously for 6 hours	1½	Colour TV	Provides 6 hours viewing	1
Electric kettle	Boils 40 cups of tea	1	VCR	Records for 10 hours	1
Coffee percolator	Makes 75 cups of coffee	1	100W bulb	Gives 10 hours illumination	1
Toaster	Toasts 70 slices of bread	1	40W fluorescent strip	Gives 20 hours illumination	1

● **Typical running costs**
The table shows you how much electricity is used, on average, by common household appliances with different kW ratings. For example, a 100W light bulb can give you 10 hours of illumination before it uses up one 1 kilowatt unit, whereas a 3 kilowatt bar fire will give off heat for only 20 minutes for the same 1 kilowatt.

UNDERSTANDING THE BASICS

Though many people imagine working on the electrical circuits of their homes to be a complicated business it is, in fact, based on very simple principles.

For any electrical appliance to work, the power must have a complete circuit, flowing along a wire from its source – a battery, for instance – to the appliance – say a light bulb – then flowing back to the source along another wire. That is a circuit, and if it is broken at any point the appliance stops working – the bulb goes out.

Breaking the circuit – and restoring it as required – is what a switch is for. With the switch on, the circuit is complete and the bulb or other appliance operates. Switching off makes a gap in the circuit so that the electricity stops flowing.

Though a break in either wire will stop the power flow, in practice a switch should be wired so that it interrupts the live wire, the one taking power to the appliance. In this way the appliance is completely dead when the switch is off. If the switch is wired so as to interrupt the neutral wire, which takes the electricity back to its source, the appliance will stop working but elements in it will still be 'live', which can be dangerous.

Though mains electricity is much more powerful than that produced by a battery it operates in exactly the same way, flowing through a live or 'phase' wire which is linked to every socket outlet, light and fixed electrical appliance in your home.

For purposes of identification when wiring is done the covering of the live wire is coloured red or brown. The covering on the neutral wire, which takes the current back out of the house after its work, is black or blue.

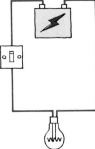

A basic circuit
Electricity runs from the source (battery) to the appliance (bulb) and returns to the source. A switch breaks the circuit to interrupt the flow of electricity.

Double insulation
A square within a square printed or moulded on an appliance means that it is double-insulated and its flex needs no earth wire.

Identifying conductors ▶
The insulation covering the conductors in cable and flex are colour-coded to indicate live, neutral and earth.

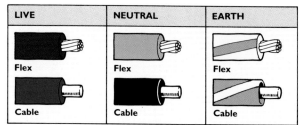

LIVE	NEUTRAL	EARTH
Flex	Flex	Flex
Cable	Cable	Cable

Earthing

Any material that electricity can flow through is known as a conductor. Most metals are good conductors of electricity, which is why metal – most often copper, which is probably the best – is used for electrical wiring.

However the earth itself, the ground on which we stand, is an extremely good conductor, a better one than the wiring used in the circuits, and associated with this is the fact that electricity will always flow into the earth if it can, and by the shortest available route. This means that if you were to touch a live conductor the current would divert and take the short route through you to the earth, perhaps with fatal consequences to you.

A similar thing can happen if a live wire comes accidentally into contact with any exposed metal component of an appliance, including its casing. To prevent this a third wire is incorporated in the wiring system and connected to the earth, usually via the outer casing of the Electricity Board's main service cable. This third, 'earth wire' is attached

to the metal casing of some appliances, and to special 'earth terminals' in others, providing a direct route to earth should a fault occur. This change of route by the power, called a 'short circuit', causes a fuse to blow or circuit breaker to operate, cutting off the current.

Some appliances are double-insulated, which usually means that they have non-conductive plastic casings that insulate the user from any metal part that could become live. For this reason double-insulated appliances do not have to be 'earthed' with a third wire.

The earth wire either has a green and yellow covering or is a bare copper wire sandwiched between the insulated live and neutral wires in the electrical cable. Whenever a bare earth wire is exposed for linking to socket outlets or lighting fittings it should be covered with a green or green-and-yellow sleeve.

Because of the danger of faulty wiring coming into contact with them, exposed metal water, heating and gas pipes should also be connected to the earthing system by a separate cable (◁).

DIY WIRING

Many householders have a certain reluctance to undertake any but the simplest jobs involving electricity, no matter how competent they may be in other areas of home improvement.

To some extent the attitude is quite justifiable. It is sensible to have a healthy respect for anything as potentially dangerous as electricity, and it would be very foolhardy of anyone to jump in at the deep end and undertake a major installation before gaining some experience on less ambitious jobs.

In the end, though, many of us are driven to doing our own house wiring by the prohibitive cost of hiring the professionals. No-one minds paying for expert knowledge, but the truth is that much of the expert's time is taken up lifting floorboards, chopping out and repairing plaster and drilling holes in walls and timbers to run the cable – all jobs that most people would be happy to do themselves.

The electrician's 'Bible'
What unnerves the householder is the possibility of making mistakes with the connections or with the choice of equipment. Fortunately we are guided in this country by a set of detailed rules laid down by the Instititution of Electrical Engineers in a document known as the IEE Wiring Regulations. This is the professional electrician's 'Bible', and it covers every aspect of electrical installation. If you follow its recommendations you can feel confident that your work will satisfy the Electricity Board. Indeed the Board will refuse to connect up an installation that doesn't comply with the Regulations.

You can buy a copy of the Wiring Regulations from the IEE itself or you can borrow one from your public library. Unfortunately the guide is notoriously difficult to understand, and it has even proved necessary to publish a 'guide to the guide', so that electricians can find their way through this exacting reference book.

The methods suggested in these pages comply with the Regulations, so you should have no need to refer to the originals unless you plan to undertake a job beyond the scope of this book.

Nevertheless, take the trouble to read all of the relevant information in the chapter so that you fully understand what you are doing. If at any time you become unsure of your competence don't hesitate to ask a professional electrician for help or advice.

FUSES AND CIRCUIT BREAKERS

A conductor will heat up if an unusually high current flows through it. This can damage electrical equipment and cause a serious risk of fire if it is allowed to continue in any part of a domestic wiring system. As a safeguard weak links are included in the wiring to break the circuit before the current can reach a dangerously high level.

The most common form of protection is a fuse, a thin wire designed to break the circuit by melting at a certain temperature depending on the part of the system it is protecting – an individual appliance, a single power or lighting circuit or a whole domestic wiring system.

Alternatively, a special automatic switch called a circuit breaker will trip and cut off the current as soon as it detects an overload on the wiring.

A fuse will 'blow' in the following circumstances:

- When too many appliances are operated on a circuit the excessive demand for current will blow the fuse in that circuit.
- When current reroutes to earth because of a faulty appliance the flow of current increases in the circuit and blows the fuse. This is called short circuiting.

WARNING: The original fault must be dealt with before the fuse is replaced.

Measuring Electricity

Watts measure the amount of power used by an appliance when working. The wattage of an electrical appliance should be marked on its casing. One thousand watts (1000W) equals one kilowatt (1kW).

Amps measure the flow of electric power necessary to produce the required wattage for an appliance.

Volts measure the 'pressure' provided by the generators of the Electricity Board that drives the current along the conductors to the various outlets. In Britain 240 volts is standard.

If you know two of these factors you can determine the other:

Watts = Amps Volts	Amps x Volts = Watts
A method to determine a safe fuse or flex.	Indicates how much power is needed to operate an appliance.

WITH SAFETY IN MIND

Throughout this chapter you will find many references to safety while actually working on any part of your electrical system, but it cannot be stressed too strongly that you must take every step to safeguard yourself and others who will later be using the system. Faulty wiring and appliances are dangerous. When you deal with electricity the rule is 'safety first'.

- Never inspect or work on any part of an electrical installation without first switching off the power at the consumer unit and removing the circuit fuse (▷).
- Disconnect any portable appliance or light fitting that is plugged into a socket before you work on it.
- Double-check all your work, especially connections, before you turn the power on again.

- Always use the correct tools for any electrical job, and use good-quality equipment and materials.
- Fuses are important safety devices. Never fit one that is rated too highly for the circuit it is to protect (▷). No other type of wire or metal strip should be used in place of proper fuses or fusewire.
- Wear rubber-soled shoes when working on an electrical installation.

Using professionals

Always be prepared to seek the advice and/or help of a professional electrician if you do not feel competent to handle a particular job yourself, especially if you discover, or even only suspect, that some part of an installation is out of date or dangerous for some other reason.

But make sure that any professional you hire is fully qualified. Check whether he or she is registered with the NICEIC (National Inspection Council for Electrical Installation Contracting). To be a member of this association an electrician must be fully cognizant of, and must comply with, the Regulations for Electrical Installations, the code of practice published by the Institution of Electrical Engineers (▷).

Electricity Board testing

Any significant rewiring, especially new circuits, must be tested by a competent electrician; and, when you apply for connection to the mains supply, you must submit a certificate to the Electricity Board stating that the rewiring complies with the Wiring Regulations. For a fee, the Board will test DIY wiring at the time of connection. Never make connections to the meter or the consumers earth terminal yourself. If you are not sure whether new wiring requires testing, contact your local Board for advice.

IS THE POWER OFF?

Having turned off the power you can make doubly sure that a particular outlet is safe to work on by plugging into it some appliance that you know to be in working order – a table lamp, for instance, which you have tested before switching off.

As a further precaution you should check on whether actual terminals or wires are live before tampering with them. Use an electronic mains tester, of the kind in the form of a screwdriver. With your fingertip on its metal cap touch the terminal or wire with the tip of the blade. An indicator in the insulated handle lights up if the terminal or wire is live. Don't use one of the cheap neon testers. Their indicators are not clear in strong light.

Using an electronic tester
With your finger on the metal cap, touch a terminal with the end of the blade. The terminal is live if the indicator glows.

BATHROOM SAFETY

Water and electricity form a very dangerous combination, for water is a highly efficient conductor of electric current. For this reason bathrooms are potentially the most dangerous areas in terms of electricity. Where there are so many exposed pipes and fittings, combined with wet conditions, stringent regulations must be observed if fatal accidents are to be avoided.

GENERAL SAFETY

● No sockets should be fitted in a bathroom except special ones approved for electric shavers (◁).

● Regulations stipulate that any standard light switches in bathrooms must be out of reach of anyone using a shower, bath or washbasin. The only sure way of complying with this is to fit nothing but ceiling-mounted pull-cord switches.

● Any bathroom heater must comply with IEE Regulations (◁).

● If you have a shower unit in a bedroom it must be at least 2.5m (8ft) from any socket outlet.

● Light fittings must also be out of reach, so fit a close-mounted ceiling light, properly enclosed, rather than a pendant fitting.

● Never use portable fires or other appliances such as hair dryers in a bathroom even if they are plugged into a socket outside.

WARNING

Have supplementary bonding tested by a qualified electrician. If you have had no previous experience of wiring and making connections have supplementary bonding installed by a professional.

Supplementary bonding

In a bathroom there are many non-electrical metallic components such as metal baths and basins, supply pipes to bath and basin taps, metal wastepipes, radiators, central heating pipework and so on, all of which could become dangerous if they were to come into contact with a live electrical conductor. To ensure that such an occurrence would blow a fuse in the consumer unit, Wiring Regulations specify that all these metal components must be connected one to another by an earth conductor which itself is connected to a terminal on the earthing block in the consumer unit. This is known as supplementary bonding and is required for new bathrooms even when there is no electrical equipment installed in the room and even though the water and gas pipes are bonded to the consumer's earth terminal near the consumer unit (◁). When electrical equipment like a heater or shower is fitted in a bathroom, that too must be supplementary bonded by connecting its metalwork, such as the casing, to the non-electrical pipework even though the appliance is connected to the earthing conductor in the supply cable.

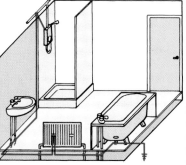

Supplementary bonding in a bathroom

Making the connections

Wiring Regulations specify the minimum size of earthing conductor that can be used for supplementary bonding in different situations so that large electrical installations can be costed economically. In a home environment use 6mm² single-core cable insulated with green or green/yellow PVC for all the supplementary bonding. It is large enough to be safe in any domestic situation unless your house is wired with Protective Multiple Earthing (◁) in which case you should consult a professional. For a neat appearance plan the route of the earth cable to run from point to point behind the bath panel, under floorboards and through basin pedestals. If necessary run the cable through a hollow wall or under plaster like any other electrical cable (◁).

Connecting to pipework

Use an earth clamp (1) to make connections to pipework. Clean the pipe locally with wire wool to make a good connection between the pipe and clamp, and scrape or strip an area of paintwork if the pipe has been painted.

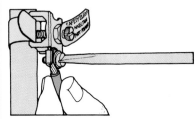

1 Fit an earth clamp to pipework

Connecting to a bath or basin

Metal baths or basins are made with an earth tag. Connect the earth cable by trapping the bared end of the conductor under a nut and bolt with metal washers (2). Make sure the tag has not been painted or enamelled.

If an old bath or basin has not been provided with an earth tag, drill a hole through the foot of the bath or to the rim at the back of the basin and connect the cable with a similar nut and bolt.

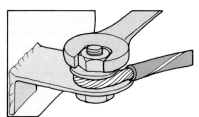

2 Connect to bath or basin earth tag

Connecting to an appliance

Connect the earth to the terminal provided in an electrical appliance (3) and run it to a clamp on a metal supply pipe nearby.

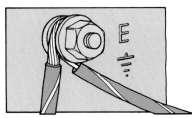

3 Fix to an appliance earth terminal

● **Supplementary bonding in a kitchen**
Supplementary bonding regulations apply to kitchens as well as bathrooms. Bond metal sink units, metallic supply and wastepipes, radiators and central heating pipework. Space and water heaters must be bonded as for bathrooms.

DEALING WITH ELECTRIC SHOCK

If someone in your presence gets an electric shock and is still in contact with its source, turn off the current at once by pulling out the plug or switching off at the socket or the consumer unit. If you cannot do this don't take hold of the person; the current may pass through you too. Pull the victim free with a dry towel, a necktie or something like that, or knock their hand free of the electrical equipment with a piece of wood. As a last resort free the victim, using their loose clothing, but without touching the body.

Don't try to move anyone who has fallen as a result of electric shock. They may have sustained other injuries. Wrap them in a blanket or coat to keep warm until they can move themselves.

Treat electrical burns by reducing the heat of the injury under slowly running cold water once the person can move and is no longer in contact with the electrical equipment. Then apply a dry dressing and seek medical advice.

Isolating the victim
If a person sustains an electric shock turn off the supply of electricity immediately, either at the consumer unit or at a socket (**1**). If this is not possible, pull the victim free with a dry towel, or knock their hand free of the electrical equipment (**2**) with a piece of wood or a broom.

ARTIFICIAL RESPIRATION

Severe electric shock can make a person stop breathing. Having freed them from the electricity supply, revive them by means of artificial respiration.

Clear the airways
Clear the victim's airways by loosening the clothing round the neck, chest and waist. Make sure that the mouth is free of food and remove any dentures (**1**). Lay the person on his or her back and tilt the head back while supporting the back of the neck with one hand.

1 Clear the mouth of food or dentures

Restart the breathing
With your free hand close and open the jaw several times in an attempt to restart the person's breathing (**2**). If this does not succeed quickly, try more direct methods of artificial respiration.

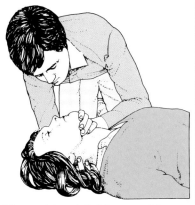

2 Open and close victim's jaw several times

Mouth to nose
Cover the victim's mouth with one hand and blow firmly into the nose (**3**). Look for signs of the chest rising and falling, then blow again three or four times in rapid succession. Repeat this procedure every four or five seconds until normal breathing resumes.

Mouth to mouth
Pinch shut the victim's nostrils, then cover the mouth with your own, making a seal all round (**4**), and proceed as in mouth to nose above.

3 Mouth to nose   **4 Mouth to mouth**

Reviving a child
With a baby or small child cover both the nose and the mouth at the same time with your own mouth (**5**), then proceed as above.

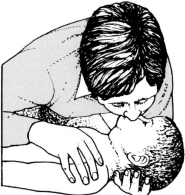

5 Cover a baby's nose and mouth

Recovery
Once breathing has started again, turn the victim face down with the head turned sideways and tilted up slightly so that the chin juts out. This will keep the airways open.

Lift one arm and one leg out from the body (**6**), then with blankets or coats arrange for the victim to stay warm while you summon medical help.

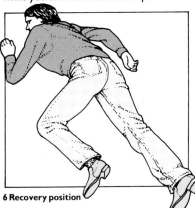

6 Recovery position

SEE ALSO

Details for: ▷
Safety tips   9

SIMPLE REPLACEMENTS

You can carry out many repairs and replacements without having to concern yourself with the wiring system installed in your home. Many light fittings and appliances are supplied with electricity through flexible cords that simply plug into the system and are easily disconnected, so there can be no risk of getting an electric shock while you are working on them.

WARNING

Never attempt to carry out electrical repairs without first disconnecting the appliance or switching off the power supply at the consumer unit (◁).

Flexible cord

All portable appliances and some of the smaller fixed ones, as well as pendant and portable light fittings, are connected to the permanent wiring system by means of conductors of flexible cord, normally called 'flex'. Each conductor in any type of flex is made up of many fine wires twisted together, and each one is insulated from the others with a covering of non-conductive material to contain the current. Insulation material is usually colour-coded to identify live, neutral and earth conductors – brown = live; blue = neutral; green/yellow = earth.

Further protection is provided on some flexible cords in the form of an outer sheathing of insulating material enclosing all the inner conductors.

Heat-resistant flex is available for enclosed light fittings and appliances whose surfaces will become hot.

COILED FLEX

A coiled flex which stretches and retracts is ideal for a portable lamp or appliance.

Coiled flex is sold as a standard length

TYPES OF ELECTRICAL FLEX

PARALLEL TWIN

Parallel twin flex has two conductors, insulated with PVC (polyvinyl chloride) and running side by side. The insulation is joined between the two conductors along the length of the flex. This flex is used mostly for low-powered appliances like shavers and some light fittings. The wires are hardly ever colour-coded.

TWISTED TWIN

Twisted twin flex is similar to parallel twin, but the PVC-insulated conductors are twisted together for extra strength to support hanging light fittings and shades. However, it is better to use a two-core sheathed flex when wiring up pendant lights. Any old rubber-insulated flex with braided cotton covering still found in some homes should be replaced.

FLAT TWIN SHEATHED

Flat twin sheathed flex has colour-coded live and neutral conductors inside a PVC sheathing. This flex is used for double-insulated light fittings and small appliances.

TWO-CORE CIRCULAR SHEATHED

This has colour-coded neutral and live conductors inside a PVC sheathing that is circular in its cross section. It is used for wiring pendant lights and some double-insulated appliances.

THREE-CORE CIRCULAR SHEATHED

This is like two-core circular sheathed flex but it also contains an insulated and colour-coded earth wire. This flex is perhaps the most commonly used for all kinds of appliances.

UNKINKABLE BRAIDED

This flex is used on appliances like kettles and irons, which are of a high wattage and whose flex must stand up to movement and wear. The three rubber-insulated conductors are strengthened by textile cords running parallel with them, all contained in a rubber sheathing bound outside with braided material.

This type of flex can be wound round the handle of a cool electric iron.

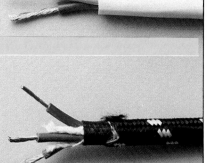

STRIPPING AND CONNECTING FLEX

Though the spacing of terminals in plugs and appliances varies, the method of stripping and connecting the flex is the same.

Stripping the flex

If the flex is sheathed slit the sheath lengthwise with a sharp knife **(1)**, being careful not to cut into the insulation covering individual conductors. Divide the conductors of parallel twin flex by pulling them apart before you expose their ends.

Peel the sheathing from the conductors, fold it back over the knife blade and cut it off **(2)**.

Separate the conductors, crop them to length and with wire strippers remove about 12mm (½in) of insulation from the end of each one **(3)**.

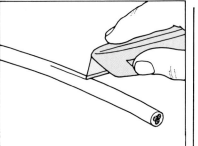

1 Slit sheathing lengthwise

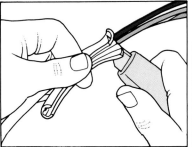

2 Fold sheathing over blade and cut it off

3 Strip insulation from conductors

MULTI-PURPOSE TOOL

A multi-purpose tool will crop and strip any size of cable or flex.

Stripping flex with a multi-purpose tool

Connecting the conductors

Twist together the individual filaments of each conductor to make them neat.

If the plug or appliance has the post type of terminals fold the bared end of wire **(1)** before pushing it in the hole. The insulation should butt against the post. Tighten the clamping screw, then pull gently on the wire to be sure it is held quite firmly.

SEE ALSO

Details for: ▷
Measuring electricity   9
Wire strippers   60

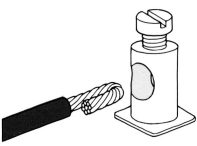

1 Post terminal

When connecting to clamp-type terminals you wrap the bared wire round the post clockwise **(2)**, then screw the clamping nut down tight on the wire and check that the conductor has been securely held.

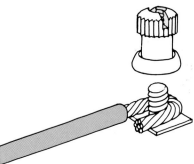

2 Clamp terminal

To attach the wire to a snap-fastening terminal swing open the back of the locking clip, insert the bared end of wire **(3)** and snap the clip back to grip the wire firmly.

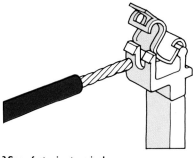

3 Snap-fastening terminal

CHOOSING A FLEX

Not only the right flex for the job is important; the size of its conductors must suit the amount of current that will be used by the appliance.

Flex is rated according to the area of the cross section of its conductors, 0.5mm² being the smallest for normal domestic wiring. Any required size is determined by the flow of current that it can handle safely. Excessive current will make a conductor overheat, so the size of the flex must be matched to the power (wattage) of the appliance which it is feeding.

Manufacturers often fit 1.25mm² flex to appliances of up to 3000W (3kW) because it is safer to use a larger conductor than necessary if a smaller flex might be easily damaged. Adopt the same procedure to replace flex.

Conductor	Current rating	Appliance
0.5 mm²	3 amps	Light fittings up to 720W
0.75 mm²	6 amps	Light fittings and appliances up to 1440W
1.0 mm²	10 amps	Appliances up to 2400W
1.25 mm²	13 amps	Appliances up to 3120W
1.5 mm²	15 amps	Appliances up to 3600W
2.5 mm²	20 amps	Appliances up to 4800W
4.0 mm²	25 amps	Appliances up to 6000W

EXTENDING FLEXIBLE CORD

When you plan the positions of socket outlets (◁) make sure there will be enough, all conveniently situated, so that it is never necessary to extend the flexible cord of a table lamp or other appliance. But if you do find that a flex will not reach a socket extend it so that it cannot be pulled tight, which can cause an accident. Never join two lengths of flex by twisting the bared ends of wires together, even if you bind them with insulating tape. People do this as a temporary measure, intending to make a proper connection later, but often forget to, and this can have fatal consequences eventually.

Flex connectors

Ideally you should fit a new length of flex, wiring it into the appliance itself. If you can't do this, or don't wish to dismantle the appliance, use a flex connector, a two- or three-terminal one, according to the type of flex.

Strip off just enough sheathing so that the conductors can reach the terminals and the sheathed part of each cord will be secured under the cord clamp at each end of the connector.

Cut the conductors to length with engineers' pliers and strip and connect them, the live conductor to one of the outer terminals, the neutral to the other and the earth wire (if present) to the central one. Make sure that matching conductors of the two cords are connected to the same terminals, then tighten the cord clamps and screw the cover in place.

In-line switches

You can extend flex by using an in-line switch instead of a flex connector. It is fitted in the same way but allows you to control the appliance from some distance away – a great advantage for the elderly or bed-ridden. Some in-line switches are fluorescent.

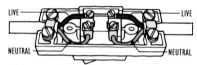

Wiring a flex conductor

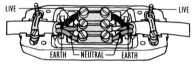

Wiring an in-line switch

Extension leads

If you fit a long flexible cord to a power tool it will inevitably become tangled and one of the conductors will eventually break, perhaps causing a short circuit. The solution is to buy or make an extension lead into which you can plug any tool or appliance you need.

The best type of extension lead to be had commercially is wound on a drum. There are 5amp ones, but it is safer to buy one rated at 13amps so that you can run a wider range of equipment with no danger of overloading it. If you use such a lead while it is wound on the drum it can overheat, so develop the habit of fully unwinding it each time.

The drums of these leads have a built-in 13amp socket to take the plug of the appliance. The plug on the lead is connected to the ordinary wall socket.

You can make an extension lead from a length of $1.5mm^2$ three-core flex with a standard 13amp plug on one end and a trailing socket on the other. Use those with unbreakable rubber casings. A trailing socket is wired similarly to a 13amp plug (See opposite). Its terminals are marked to indicate which conductors connect to them.

'Multi-way' trailing sockets will take more than one plug and are ideal for hi-fi systems with individual components that must be connected to mains supply. With a multi-way socket in the cabinet the whole system is supplied from one plug in the wall socket.

Alternatively you can use a lightweight two-part flex connector. One half has three pins which the other half receives.

Unwind a lead
Always fully unwind a 13amp extension lead before you plug in an appliance rated at 1.4kW or more.

WARNING

When wiring a two-part flex connector never attach the part with the pins to the extension lead. The exposed pins will become live – and dangerous – when the lead is plugged into the socket. In fact nothing electrical should ever be wired so that a plug can become live other than when its pins are concealed in a socket.

TYPES OF FLEX EXTENDER

Below are illustrated four of the devices available for extending the flexible cords of electrical appliances (See left).

Drum-type extension lead

13amp plug and trailing socket

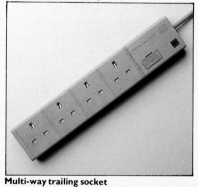

Multi-way trailing socket

Two-part flex connector

WIRING A PLUG

In the past there were many types of plug. Nowadays there are standard ones for all light fittings and portable appliances – 13amp square-pin plugs. They are available with rigid plastic or unbreakable rubber casings. Some have integral neon indicators to show when they are live, and pins insulated for part of their length now prevent the user getting a shock from a plug pulled partly from the socket.

Safety standards and fuses

Use only plugs marked BS 1363, which means conformity to British Standards and therefore safety in use. All square-pin plugs have a small cartridge fuse to protect the appliance – 3amp (red) for appliances of up to 720W or 13amp (brown) for those of from 720 to 3000W. There are 2, 5 and 10 amp fuses but they are rarely used in the home.

Wiring a 13amp plug

Loosen the large screw between the pins and remove the cover. Position the flex on the open plug to determine how much sheathing to remove, remembering that the cord clamp must grip sheathed flex, not the conductors.

Strip the sheathing and again position the flex on the plug so that you can cut the conductors to the right length.

These should take the most direct routes to their terminals and lie neatly in the channels of the plug.

Strip and prepare the ends of the wires, then secure each to its terminal. If you are using two-core flex, wire to the live and neutral terminals, leaving the earth terminal empty.

Tighten the cord clamp to grip the end of the sheathing and secure the flex. One type of plug has a sprung cord grip that tightens if the flex is pulled hard.

Check that a fuse of the correct rating is fitted.

Replace the plug's cover and tighten up the screw.

Wiring older plugs

If your home still has old round-pin sockets you must go on using round-pin plugs, which are not fused. Use the small 2amp one for lighting only, a 5amp one for appliances of up to 1000W and a 15amp one for anything between 1000W and 3000W. You should have your wiring upgraded as soon as possible so that you can use modern fused square-pin plugs.

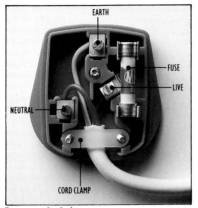

Post-terminal plug

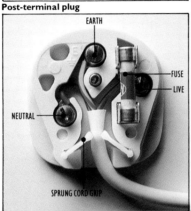

Clamp-terminal plug

Snap-fastening terminal plug

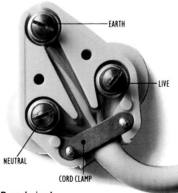

Round-pin plug

REPLACING A PENDANT LAMPHOLDER

Because they are not easy to inspect damaged lampholders can go unnoticed. You should check their condition every so often and replace any suspect ones before they become dangerous.

Pendant lampholders, which hang on flex from the ceiling, are in a stream of hot air rising from the bulb, and in time this can make plastic holders brittle and more easily cracked or broken. On a metal lampholder the earth wire can become detached or corroded so that the fitting is no longer safe.

Types of lampholder

Plastic lampholders are the most common. These have a threaded skirt that screws onto the actual holder, the part that takes the bulb, and some versions have an extended skirt for fitting in bathrooms. You should fit heat-resistant plastic holders if you use a close-fitting or badly ventilated shade.

Plastic holders are designed to take only two-core flex. Don't fit one on a three-core flex as it will have no place to attach the earth wire.

Metal lampholders are similar in their construction but they must be wired with three-core flex so that they can be connected to earth. Never fit a metal lampholder in a bathroom, and never attach one to a two-core flex, which has no earth conductor.

Fitting a lampholder

Before you start remove the fuse for the circuit or switch off the circuit breaker at the consumer unit (▷) so that no-one can turn the power on.

Unscrew the old holder's cap – or the retaining ring if it's a metal one – and slide it up the flex to expose the terminals. Loosen their screws and pull the wires out. If some wires are broken or brittle cut back slightly to expose sound wire before fitting the holder.

Slide the cap of the new fitting up the flex and attach it temporarily with adhesive tape.

Fit the live or neutral wire into either terminal and twist the conductors round the supporting lugs of the holder to take the weight off the terminals, then screw the cap down.

On a metal holder pass the earth wire through the hole in the cap before you screw it down. Wrap the wire clockwise round the fixing screw and tighten it. Screw down the retaining ring to secure the cap.

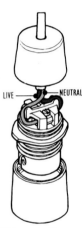

Wiring a plastic pendant lampholder

Wiring a metal pendant lampholder

15

MAIN SWITCH EQUIPMENT

Electricity flows because of a difference in pressure between the live wire and the neutral one, and this difference in the pressures is measured in volts.

Domestic electricity is supplied at 240 volts (240V) 'alternating current' by way of the Electricity Board's service cable, which enters your house underground, though in some areas power is distributed by overhead cables.

The sealing chamber

The main cable terminates at the service head, or sealing chamber, which contains the service fuse. This fuse prevents the neighbourhood being affected if there should be a serious fault in your circuitry. A cable connects the sealing chamber to the meter, which registers how much power is used. The meter and sealing chamber belong to the Electricity Board and must not be tampered with. The meter is sealed to detect interference.

If you use the cheap night-time power for storage heaters and hot water a time switch will be mounted somewhere between the sealing chamber and the meter.

Consumer units

Electricity is fed to and from the consumer unit by 'meter leads', thick single-core cables made up of several wires twisted together. The consumer unit is a box that contains the fuseways which protect the individual circuits in the house. It also incorporates the main isolating switch with which you can cut off the supply of power to the whole of the house.

In a house where several new circuits have been installed over the years the number of circuits may exceed the number of fuseways in the consumer unit, so an individual switchfuse unit – or more than one – may have to be mounted alongside the main unit. Switchfuse units comprise a single fuseway and an isolating switch. They too are connected to the meter by means of meter leads.

If your home is heated by storage heaters you will probably have a separate consumer unit for the circuits that supply the heaters.

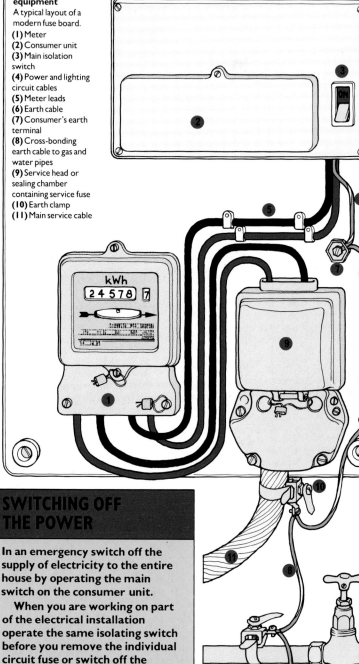

Main switch equipment
A typical layout of a modern fuse board.
(1) Meter
(2) Consumer unit
(3) Main isolation switch
(4) Power and lighting circuit cables
(5) Meter leads
(6) Earth cable
(7) Consumer's earth terminal
(8) Cross-bonding earth cable to gas and water pipes
(9) Service head or sealing chamber containing service fuse
(10) Earth clamp
(11) Main service cable

SWITCHING OFF THE POWER

In an emergency switch off the supply of electricity to the entire house by operating the main switch on the consumer unit.

When you are working on part of the electrical installation operate the same isolating switch before you remove the individual circuit fuse or switch off the miniature circuit breaker which will cut the power to the relevant circuit (◁). That circuit will be safe to work on even when you restore the power to the rest of the house with the main isolating switch.

EARTHING SYSTEMS

The earthing system

All of the individual earth conductors of the various circuits in the house are connected to one heavy earth cable in the consumer unit. This cable is sheathed in green or green/yellow and runs from the unit to the consumer's earth terminal. In most town houses the earth cable continues from the earth terminal to a clamp on the metal sheath of the main service cable just below the sealing chamber. This is an effective path to earth. The current will pass along the sheath to the Electricity Board's substation, where it is solidly connected to earth.

Until recently most electrical installations were earthed to the cold water supply so that earth-leakage current passed out along the metal water pipes into the ground in which they were buried. But more and more water systems now use non-metallic, non-conductive pipes, so that means of earthing is no longer reliable. Despite this you will find that your pipework is connected to the earth terminal in case one of the live conductors in the house should touch a pipe at some point. The same earth cable is usually clamped to a nearby gas pipe on the house side of the meter before running on to the consumer's earth terminal. This ensures that both water and gas piping systems are cross-bonded so that earth-leakage current passing through either system will run without hindrance to the clamp on the service cable sheath and so to earth. The clamps must never be interfered with.

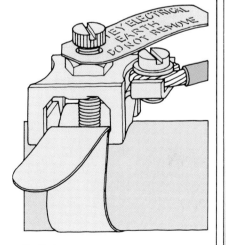

Earth clamp
The earthing system of the house finally connects to the main service cable with this type of clamp. It should not be removed under any circumstances.

PME

Sometimes, especially in country areas, the Electricity Board provides another method of earthing the system called 'protective multiple earth' (PME) by which earth-leakage current is fed back to the substation along the neutral return wire, and so to earth.

Regulations regarding the earthing of the system are particularly stringent, so if you live in a house with PME (check this with your Electricity Board) you should engage a qualified professional electrician for any but the most minor type of work.

RCCBs

Though the Electricity Board normally provides effective earthing, actually it is the consumer's responsibility. It is often achieved by installing a 'residual-current circuit breaker' or RCCB, into the house circuitry.

Under normal conditions the current flowing out through the neutral conductor is exactly the same as that flowing in through the live one. Should there be an imbalance between the two caused by an earth leakage the RCCB will detect it immediately and isolate the system.

An RCCB can be either installed as a separate unit or incorporated into the consumer unit together with the main isolating switch.

A separate unit containing an RCCB

RECOGNIZING AN OLD FUSE BOARD

Domestic wiring was once very different from a modern system. Beside lighting, water-heating and cooker circuits, each individual socket outlet had its own circuit and fuse, while further circuits would usually be installed from time to time as the needs of the household changed. Consequently an old house may have a mixture of 'fuse boxes' attached to a fuse board along with the meter. The wiring itself may be haphazard and badly labelled, with the constant risk that you have not safely isolated the circuit you wish to work on. Further, you will be unable to tell if a given fuse is correctly and safely rated unless you know what type of circuit it is protecting.

If your home still has such an old-style fuse board have it inspected and tested by a qualified electrician before you attempt to work on any part of the system. He can advise you on whether to replace the installation with a modern consumer unit or, if the system is in good working condition, he can at least label the various circuits clearly to help you in the future.

An old-fashioned fuse board
This type of installation is out of date. A professional electrician may advise you to replace at least some of the components.

● **RCCB**
An RCCB may be referred to as an RCD – residual-current device. It was formerly known as an ELCB – earth-leakage circuit breaker.

THE CONSUMER UNIT

The consumer unit is the heart of your electrical installation, for every circuit in the house must pass through it. There are several different types and styles of consumer unit but they are all based on similar principles.

Every unit has a large main switch that can turn off the whole installation. On some, more expensive units the switch is in the form of an RCCB (◁) which can be operated manually but will also 'trip' automatically, isolating the entire household system, should any serious fault occur, and much more quickly than the Electricity Board's fuse would take to blow in a similar emergency.

Some consumer units are designed so that it is impossible to remove the outer cover without first turning off the main switch. Even if yours is not of this type you should always switch off before exposing any of the elements within the consumer unit.

Having turned off the main switch, remove the cover or covers so that you can see how the unit is arranged. The cover must be replaced before the unit is switched on again – and remember: even when the unit is switched off the cable connecting the meter to the main switch is still live, so take care.

Take note of the cables that feed the various circuits in the house. Ideally they should all enter the consumer unit from the same direction.

The black-insulated neutral wires run to a common neutral block where they are attached to their individual terminals. Similarly the green earth wires run to a common earthing block. The red-covered live conductors are connected to terminals on individual fuseways or circuit breakers.

Some wires will be twisted together in one terminal. These are the two ends of a ring circuit (◁) and that is how they should be wired.

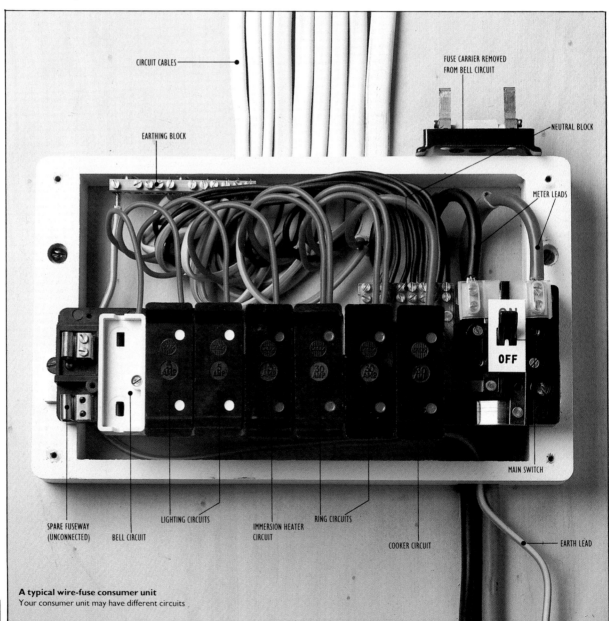

CIRCUIT CABLES

FUSE CARRIER REMOVED FROM BELL CIRCUIT

EARTHING BLOCK

NEUTRAL BLOCK

METER LEADS

ON OFF

MAIN SWITCH

SPARE FUSEWAY (UNCONNECTED)

BELL CIRCUIT

LIGHTING CIRCUITS

IMMERSION HEATER CIRCUIT

RING CIRCUITS

COOKER CIRCUIT

EARTH LEAD

A typical wire-fuse consumer unit
Your consumer unit may have different circuits

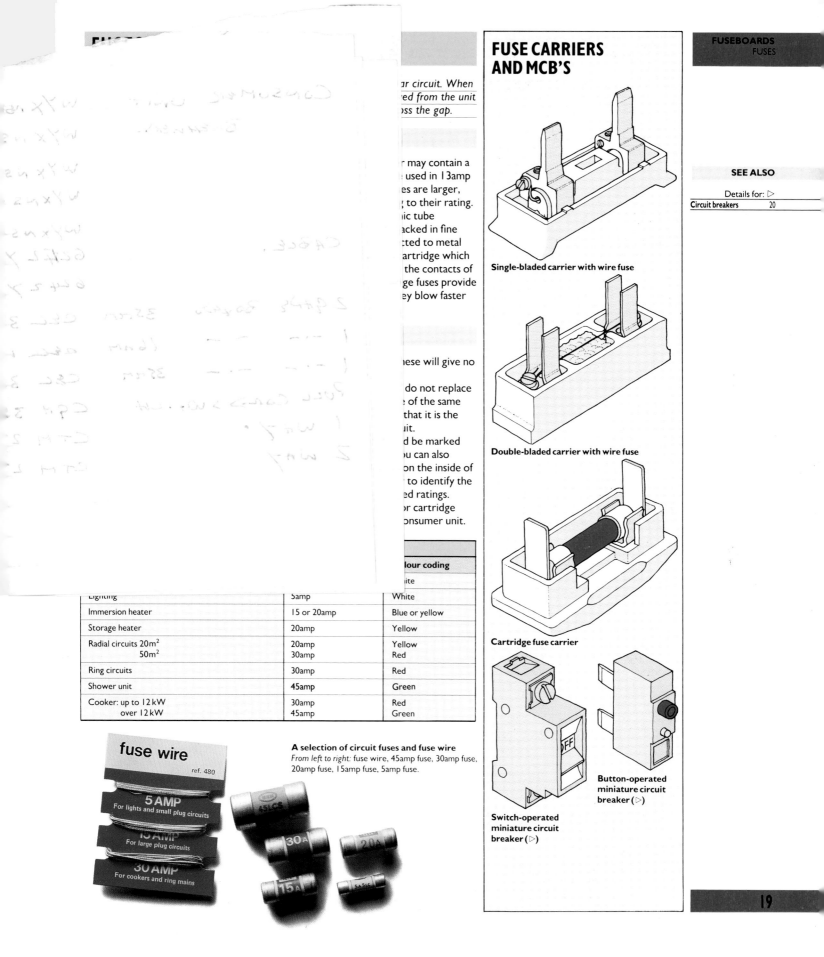

FUSE CARRIERS AND MCB'S

Single-bladed carrier with wire fuse

Double-bladed carrier with wire fuse

Cartridge fuse carrier

Switch-operated miniature circuit breaker (▷)

Button-operated miniature circuit breaker (▷)

SEE ALSO

Details for: ▷
Circuit breakers   20

*...ar circuit. When
...ed from the unit
...oss the gap.*

...r may contain a
...used in 13amp
...es are larger,
...to their rating.
...ic tube
...acked in fine
...ted to metal
...artridge which
...the contacts of
...ge fuses provide
...ey blow faster

...ese will give no

...do not replace
...e of the same
...that it is the
...it.
...d be marked
...u can also
...on the inside of
...to identify the
...ed ratings.
...or cartridge
...onsumer unit.

		Colour coding
Lighting	5amp	White
Immersion heater	15 or 20amp	Blue or yellow
Storage heater	20amp	Yellow
Radial circuits 20m²	20amp	Yellow
50m²	30amp	Red
Ring circuits	30amp	Red
Shower unit	45amp	Green
Cooker: up to 12 kW	30amp	Red
over 12 kW	45amp	Green

fuse wire
ref. 480

5 AMP
For lights and small plug circuits

15 AMP
For large plug circuits

30 AMP
For cookers and ring mains

A selection of circuit fuses and fuse wire
From left to right: fuse wire, 45amp fuse, 30amp fuse, 20amp fuse, 15amp fuse, 5amp fuse.

CHANGING A FUSE

When everything on a circuit stops working your first step is to check the fuse on the circuit and see if it has blown. Turn off the main switch on the consumer unit, remove the cover and look for the failed fuse. The fuse will be easier to find if you know which circuit is affected, so check the list of circuits inside the unit's cover. If there is no list you will have to inspect all likely circuits. For instance, if the lights 'blew' when you switched them on you need check only the lighting circuits. These are colour-coded white.

Checking a cartridge fuse

The simple way to check a suspect cartridge fuse is to replace it with a new one and see if the circuit works. Or you can check the fuse with a metal-cased torch. Remove the bottom cap of the torch and touch one end of the fuse to the base of the battery while resting its other end against the battery's metal casing. If the battery bulb lights up the fuse is sound.

Testing a cartridge fuse
With the torch switched on, hold the fuse against the battery and the metal casing.

Checking a rewirable fuse

On a blown rewirable fuse a visual check will usually detect the broken wire and scorch marks on the fuse carrier. If the fuse is one on which you cannot see the whole length of the fuse wire you should pull gently on each end of the wire with the tip of a screwdriver to see if it is intact.

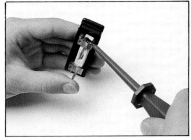

Pull the wire gently with a small screwdriver

HOW TO REPLACE FUSE WIRE

To replace a blown fuse wire loosen the two terminals holding the fuse and extract the broken pieces. Wrap one end of a new length of wire clockwise round one terminal and tighten the screw on it **(1)**, then run the wire across to the other terminal, leaving it slightly slack, attach it in the same way **(2)** and cut off any excess from the ends.

If the wire passes through a tube in the fuse carrier it has to be inserted before either terminal is tightened **(3)**.

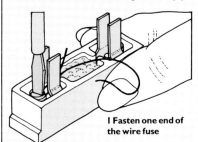

1 Fasten one end of the wire fuse

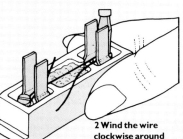

2 Wind the wire clockwise around the other terminal

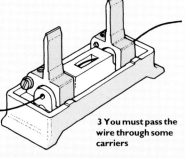

3 You must pass the wire through some carriers

IF THE FUSE BLOWS AGAIN

If a replaced fuse blows again as soon as the power is switched on there is a fault or an overload – too many appliances plugged in – on that circuit and it must be detected and rectified before another fuse is inserted.

Circuit breakers

In some consumer units you will find miniature circuit breakers (MCBs) instead of the usual fuse holders. These are amp-rated just like fuses but instead of removing an MCB to isolate the circuit you merely operate a switch or a button on it to 'off'. When a fault occurs the circuit breaker switches to the 'off' position automatically, so the faulty circuit is obvious as soon as you open the consumer unit. Turn the main switch off, then simply close the switch on the MCB to reset it. There is no fuse to replace. If the switch or button will not stay in the 'on' position when power is restored there is still a fault on the circuit which must be rectified.

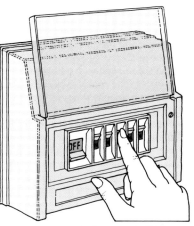

With the main switch off, turn on the MCB

Checking out a fault

An electrician can test a circuit for you with special equipment, but first carry out some simple tests yourself.

Before inspecting any part of the circuit turn off the consumer unit's main switch, remove the relevant fuse holder and keep it in your pocket so that no one can replace it while you work. If you have circuit breakers instead of fuses fix the switch at 'off' with bright adhesive tape and leave a note on the consumer unit.

Unplug all appliances on the faulty circuit to make sure that it is not simply overloaded, then switch on again. If the circuit is still faulty switch off again and inspect the socket outlets and light fittings to see if a conductor has worked loose and is touching one of the other wires, terminals or outer casing, causing a short circuit.

If none of this enables you to find the fault call in an electrician.

TYPES OF DOMESTIC ELECTRICAL CIRCUITS

Running from the consumer unit are the cables which supply the various fixed wiring circuits in your home. Not only are the sizes of the cables different (▷); the circuits themselves also differ, depending on what they are used for and also, in some cases, how old they happen to be.

RING CIRCUITS

The most common form of 'power' circuit for feeding socket outlets is the ring circuit, or 'ring main', in which a cable starts from terminals in the consumer unit and goes right round the house, connecting socket to socket, and arriving back at the same terminals. By this method power can reach any of the socket outlets or fused connection units (▷) from both directions, which reduces the load on the cable.

Ring mains are always run in 2.5mm^2 cable and are protected by 30amp fuses. There is no limit to the number of socket outlets or fused connection units that can be fitted to one ring circuit so long as it does not serve a floor area of more than 100sq m (120sq yd), a limit based on the number of electric heaters which would be adequate to heat that space. In practice most two-storey dwellings have one ring main for the upper floor and another for downstairs.

Spurs
The number of sockets on a ring main can be increased by adding extensions, or 'spurs'. A spur can be a single 2.5mm^2 cable connected to the cables of an existing socket or fused connection unit or it can run from a junction box inserted in the ring.

Current regulations allow each spur to serve one fused connection unit for a fixed appliance or one single or double socket outlet. You can have as many spurs on a ring circuit as there were sockets on it originally, and for this calculation a double socket is counted as two. The 30amp fuse that protects the ring main is unchanged no matter how many spurs are connected to the circuit.

RADIAL CIRCUITS

A radial power circuit feeds a number of socket outlets or fused connection units but, unlike a ring circuit, its cable terminates at the last outlet. The size of cable and the fuse rating depend on size of the floor area to be supplied by the circuit. In a room of up to 20sq m (24sq yd) the cable should be 2.5mm^2, protected by a 20amp MCB or fuse of any type. For a larger area, up to 50sq m (60sq yd) you should use 4mm^2 cable with a 30amp cartridge fuse or MCB, but a rewirable fuse is not permitted.

Any number of socket outlets can be supplied by one of these circuits, and spurs can be added if required. The circuits are known as multi-outlet radial circuits, but a powerful appliance like a cooker or shower unit must have its own radial circuit (▷).

LIGHTING CIRCUITS

Domestic lighting circuits are also of the radial kind, but there are two systems currently in use.

The loop-in system has a single cable that runs from ceiling rose to ceiling rose, terminating at the last one on the circuit while single cables run from the ceiling roses to the light switches.

The older, junction-box system incorporates a junction box for each light. The junction boxes are situated conveniently on the single supply cable. A cable runs from each junction box to the ceiling rose and another from the box to the light switch. In practice, most lighting systems are combinations of the two methods.

A single circuit of 1mm^2 cable can serve the equivalent of twelve 100W light fittings, though you might have to reduce the number if you wished to install multi-light fittings or if one of your lamps was more powerful than 100W. Once again it is more practical to have two or more separate lighting circuits running from the consumer unit.

Lighting circuits are protected by 5amp fuses.

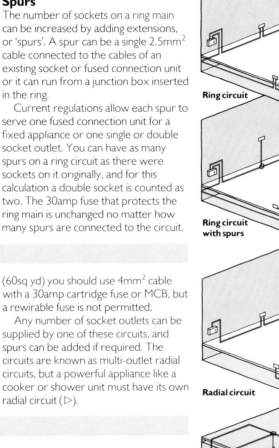

Ring circuit

Ring circuit with spurs

Radial circuit

Loop-in system

Junction-box system

21

CABLE: TYPES

Two-core and earth

Cable for the fixed wiring of electrical systems normally has three conductors: the insulated live and neutral ones and the earth conductor lying between them, uninsulated except for the sheathing that encloses all three. Cable up to 2.5mm^2 has solid, single-core conductors, but larger sizes, up to 10mm^2, would be too stiff with solid conductors so each one is made up of seven strands. The live conductor is insulated with red PVC and the neutral one with black. When an earth conductor is exposed, as in a socket outlet, it should be covered with a green and yellow sleeve. You can buy this from any electricians' supplier. The PVC sheathing on the outside of the cable is usually white or grey.

Heat-resistant cable is available for use in a situation where extra heat may be generated, and there is heat-resistant sleeving for the conductors in enclosed light fittings.

Three-core and earth

This type of cable is used in two-way lighting systems, which can be switched on and off at different switches. It contains three insulated conductors and a bare earth wire. The conductors have red, yellow and blue coverings.

Single-core cable

Insulated single-core cable is used in buildings where the electrical wiring is run in metal or plastic conduit. This is rare in domestic buildings. The cable is colour-coded in the normal way: red for live, black for neutral and green/yellow for earth.

Single-core 16mm^2 cable is used for connecting the consumer unit to the earth, and single-core 16mm^2 cable is used for connecting the unit to the meter. Meter leads are insulated and sheathed in red for the live conductor and black for the neutral one.

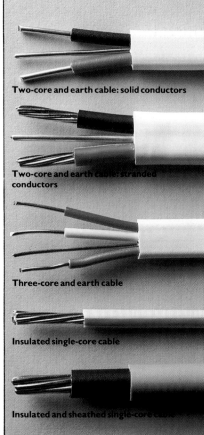

Two-core and earth cable: solid conductors

Two-core and earth cable: stranded conductors

Three-core and earth cable

Insulated single-core cable

Insulated and sheathed single-core cable

OLD CABLE

Houses which were wired before World War II may still have old rubber-sheathed and -insulated cable, and some old cable may even be sheathed in lead.

Rubber sheathing is usually a matt black. It is more flexible than the modern PVC insulation unless it has deteriorated, when it will be crumbly.

This type of cable may be dangerous

CIRCUIT CABLE SIZES		
Circuit	**Size**	**Type**
Lighting	1·0mm^2	Two-core and earth
Bell or chime transformer	1·0mm^2	Two-core and earth
Immersion heater	2·5mm^2	Two-core and earth
Storage heater	2·5mm^2	Two-core and earth
Ring circuit	2·5mm^2	Two-core and earth
Spurs	2.5mm^2	Two-core and earth
Radial — 20 amp	2·5mm^2	Two-core and earth
Radial — 30 amp	4·0mm^2	Two-core and earth
Shower unit	10.0mm^2	Two-core and earth
Cooker up to 12 kW	6·0mm^2	Two-core and earth
Cooker over 12 kW	10·0mm^2	Two-core and earth
Consumer earth cable	16.0mm^2	Single core
Meter leads	16·0mm^2	Single core
	All cable sizes in square millimetres	

STRIPPING CABLE

When cable is wired up to an accessory some of the sheathing and insulation must be removed.

Slit the sheathing lengthwise with a sharp knife, peel it off the conductors, fold it over the blade and cut it off.

Take about 12mm (½in) of insulation off the ends of the conductors using wire strippers.

Cover the uninsulated earth wire with a green/yellow sleeve, leaving 12mm (½in) of the wire exposed for connecting to the earth terminal.

If more than one conductor is to be inserted in the same terminal twist the exposed ends together with strong pliers to ensure the maximum contact for all of the wires.

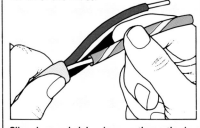

Slip colour-coded sleeving over the earth wire

INSIDE A HOLLOW WALL

For a short cable run on a lath-and-plaster wall, hack the plaster away, fix the cable to the studs and then plaster over again in the normal way.

While you can run cable through the space between the two claddings of a stud partition wall there is no way of doing this without some damage to the wall and the decoration. Drill a 12mm (½in) hole through the top wall plate above the position of the switch, then tap the wall directly below the hole to locate the nogging (▷). Cut a hole in the lath and plaster to reveal the top of the nogging and drill a similar hole through it.

Pass a lead weight on a plumb line through both of the holes and down to the location of the switch. Tie the cable to the line and pull it through.

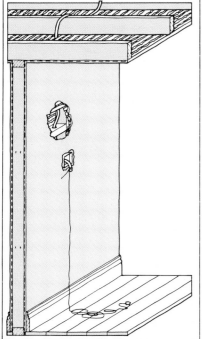

Running cable through a hollow wall
If a nogging prevents you running cable directly to a switch, cut away some laths and plaster to drill a hole through the timber.

RUNNING CABLE

Long runs of cable are necessary to take power from the consumer unit to all the sockets, light fittings and fixed appliances in the home.

The cable must be fixed securely to the structure of the house along its route except in confined spaces to which there is normally no access, such as voids between floors and inside hollow walls. There are accepted ways of running and fixing cable, depending on particular circumstances.

Surface fixing

PVC-insulated and -sheathed cable can be fixed directly to the surface of a wall or ceiling without any further protection. Fix it with plastic cable clips **(1)** or metal buckle clips **(2)** every 400mm (1ft 4in) on vertical runs and every 250mm (10in) on horizontal runs. Keep the runs as straight and neat as possible, and when several cables run in the same direction group them together. Avoid kinks in the cable by keeping it on the drum as long as possible, but if you do have to get any kinks out pull the cable round a thick dowel held in a vice.

If a cable seems vulnerable you can cover it with an impact-resistant plastic channel **(3)**. Having secured the cable with clips, you simply nail the channel in place over it.

1 Plastic cable clip

2 Metal buckle clip

3 Impact-resistant plastic channel

Concealed fixing

While surface-fixed cable is quite acceptable in cellars, under stairs and in workshops and garages, few people want to see it running across their living room walls or ceilings. It's better to bury it in the plaster or hide it in a wall void. Sheathed cable can be buried without further protection.

Wherever possible cable should run vertically to switches or sockets, to avoid dangerous clashes with wall fittings or fixtures installed later. Most people allow for it to run this way. If you must run horizontal cable confine it to within 150mm (6in) of the ceiling or 300mm (1ft) of the floor. Never run a buried cable diagonally across a wall.

Some people cover all buried cable with a channel, but it isn't required by the IEE Regulations.

Cable buried in light plastic conduit can be withdrawn later, if necessary, without disturbing decorations, but the need is so rare in a house as to be hardly worth considering.

Mark out your cable runs on the plaster, allowing a channel about 25mm (1in) wide for single cable. Cut both sides with a bolster and club hammer and hack out the plaster between the cuts with a cold chisel. Normally plaster is thick enough to conceal cable, but you may have to chop out some brickwork to get the depth. Clip the cable in the channel **(1)** and, when you have checked that the installation is working, plaster over it. To avoid electric shock ensure that the power to that circuit is turned off before you use wet plaster round a switch or socket outlet **(2)**.

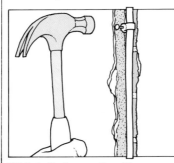

1 Nail plastic clips over the cable

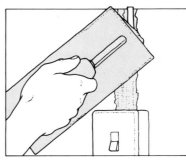

2 Repair the plaster up to the switch

RUNNING CABLE UNDER FLOORS

Power and lighting circuits are often concealed beneath floors if access is possible. It isn't necessary to lift every floorboard to run a cable from one side of a room to the other; by lifting a board every 2m (6ft) or so you should be able to pass the cable from one gap to the next with the help of a length of stiff wire bent into a hook at one end. Look for boards that have been taken up before, as they will be fairly easy to lift and you will damage fewer boards.

Lifting floorboards

Lifting square-edged boards

Drive a wide bolster chisel between two boards about 50mm (2in) from the cut end of one of them (1). Lever that board up with the bolster, then do the same on its other edge, working along the board until you have raised it enough to wedge a cold chisel under it (2). Proceed along the board, raising it with the chisel, until the board is loose.

Full-length boards

If you have to lift a board that runs the whole length of the floor, from one skirting to the other, start somewhere near the middle of the board and close to a floor joist. The nail heads indicate the positions of joists. Lever the board up and make a saw cut through it centred on the joist, then lift the board in the normal way.

Lifting tongued and grooved boards

You cannot lift a tongued and grooved floorboard until you have cut through the tongues along both sides of the board, either with a special floorboard saw, which has a blade with a rounded tip, or with a power jigsaw.

Cutting a full-length board
Cut a full-length board in two directly over a floor joist.

● **Removing skirting**
Remove skirting by levering it away from the wall with a crowbar or bolster chisel. Protect the plaster above the skirting with a slim block of wood. Work along the skirting little by little until the fixing nails loosen.

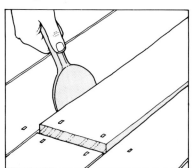

I Prise up the floorboard with a bolster

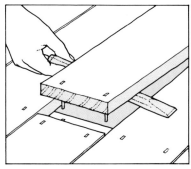

2 Wedge the raised end with a cold chisel

CUTTING A BOARD NEXT TO A SKIRTING

Should you need to cut through a board that lies close to a wall it may not be possible to lift it without damaging the bottom edge of the skirting board. In such a case drill a starting hole through the board alongside the joist nearest to the wall, insert the blade of a padsaw or power jigsaw in the hole and cut across the board flush with the side of the joist (1). To support the cut end afterwards nail a length of 50 x 50mm (2 x 2in) softwood to the joist. Hold it up tightly against the undersides of the adjacent floorboards while fixing it to ensure that the cut board will lie flush with the others (2).

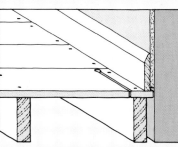

I Cut through a trapped board with a jigsaw

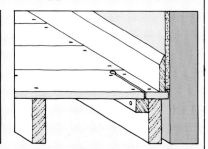

2 Support the cut board with a nailed batten

Solid floors

In a new concrete floor you can lay conduit and run cable through it before the concrete is poured.

In an existing solid floor you can cut a channel for conduit – hard work without an electric hammer and chisel bit – but if the floor is tiled you will not want to spoil it for one or two socket outlets. An alternative is to drop spur cables, buried in the wall plaster, from the ring circuit in the upper floor. Another way is to run cable through the wall from an adjacent area and channel it horizontally in the plaster just above the skirting. Yet another is to remove the skirting (◁), clip the cable to the wall and cover it with protective channel. Note the position of the cable to avoid piercing it when you nail back the skirting board.

In the roof space

In the roof space all wiring can be surface-run, but as people may enter it occasionally you must see that the cable is clipped securely to the joists or rafters. You can even run cable along the tops of joists in some areas, but run it through holes in the normal way where joists are to be boarded over or in areas of access such as round water tanks and the entrance hatch itself.

Wiring overlaid by roof insulation has a slightly higher chance of heating up. Ring mains and lighting circuits do not present a problem , but circuits to heaters, cookers or shower units are more critical. Wherever possible run cable over thermal insulation. If you cannot avoid running it under the material you should use a heavier cable to be on the safe side.

When expanded-polystyrene insulation is in contact with electric cable for a long time it affects the plasticizer in the PVC sheathing on the cable. The plasticizer moves to the surface of the sheathing, reacts with the polystyrene and forms a sticky substance on the cable. This becomes a dry crust which cracks if the cable is lifted out of the roof insulation and bent. It gives the impression that the cable insulation is cracking, but scientific testing has shown that the cracking is merely in the surface crust. On balance it is best to keep cable away from polystyrene.

RUNNING CABLE

Running cable through the house structure
Use the most convenient method to run cable to sockets and switches.
1 Clip cable to roof timbers in loft.
2 Junction boxes must be fixed securely.
3 Run cable through holes in the joists near the hatch.
4 Run cable over loft insulation.
5 To avoid damaging a finished floor you can run a short spur through the wall from the next room.
6 When cable runs across the line of joists drill holes 50mm (2in) below the top edges.
7 Cable running parallel to joists can lay on the ceiling below.
8 Let cable drape onto the base below a suspended floor.
9 When you cannot run cable through a concrete floor you can drop a spur from the floor above.
10 Take the opportunity to bury conduit for cable in a new concrete floor.

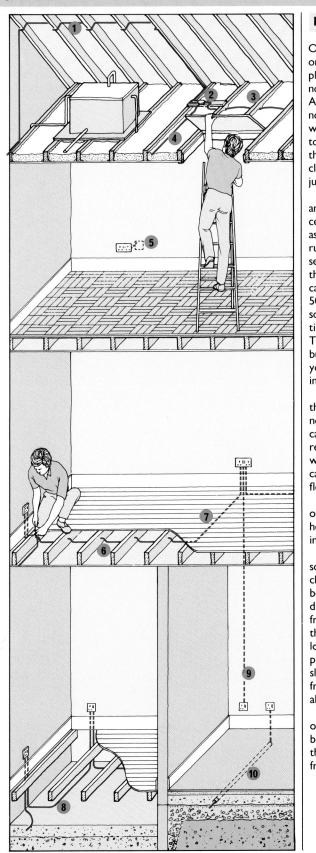

Running the cable

On the ground floor the cable can rest on the earth or on the concrete platform below the joists if there will not normally be access to the space. Allow enough slack so that the cable is not suspended above the platform, which might put a strain on any fixings to junction boxes or socket outlets. For the same reason secure the cable with clips to the side of the joist beside junction boxes or other accessories.

When laying cable between a floor and the ceiling below, it can rest on the ceiling without any other fixing so long as it runs parallel with the joists. If it runs at right angles to the joists drill a series of 12mm ($\frac{1}{2}$in) holes, one through each joist along the intended cable run. The holes must be at least 50mm (2in) below the tops of the joists so that floorboard nails cannot at some time be hammered through the cable. The space between the joists is limited but you can hire a special joist brace or you can cut down a spade bit and use it in a power drill.

When you reach the last joist against the wall, instead of drilling a hole cut a notch in its top so that you can feed the cable up behind the skirting board to reach a socket outlet. Cover the notch with a stout metal plate to protect the cable, then cut a notch in the end of the floorboard to clear it.

Never attach electrical cable to gas or water pipes, and don't run it next to heating pipes, as the heat can melt the insulation.

Having marked out the position of a socket or fused connection unit, cut a channel from it down to the skirting board and, with an extra-long masonry drill in a power tool, remove the plaster from behind the skirting board. By using the drill at a shallow angle you can loosen much of the debris, but you will probably have to finish the job with a slim cold chisel. Raking the debris out from below with the same chisel will also help to dislodge it.

Pass a stiff wire with a hook formed on its end down behind the skirting board, hook the cable and pull it through, at the same time feeding it from below with your other hand.

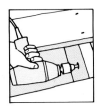

Drilling the joists
Shorten a spade bit so that your drill fits between the joists.

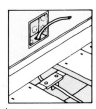

Notching a joist
Screw a thick metal plate over a notch in a joist close to a wall.

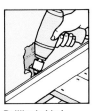

Drilling behind skirting
Use a long masonry drill to remove the plaster behind a skirting board.

ASSESSING YOUR INSTALLATION

Inspect your electrical system to ensure that it is safe and adequate for your future needs. But remember, you should never examine any part of it without first switching off the power at the consumer unit (◁).

If you are in doubt about some aspect of the installation you should ask a qualified electrician his opinion. If you get in touch with your Electricity Board they will arrange for someone to test the whole system for you.

QUESTIONS	ANSWERS
Do you have a modern consumer unit or a mixture of old 'fuse boxes' (◁)?	Old fuse boxes can be unsafe and should be replaced with a modern unit. Seek professional advice on this.
Is the consumer unit in good condition?	Replace broken casing or cracked covers and check that all fuse carriers are intact and that they fit snugly in the fuseways.
Are the fuse carriers for the circuits clearly labelled?	If you cannot identify the various circuits, have an electrician test the system and label the fuses.
Are all your circuit fuses of the correct ratings (◁)?	Replace any fuses of the wrong rating. If an unusually large fuse is protecting one of the circuits get professional advice before changing it. It may have a special purpose. Any wire other than proper fuse wire found in a fuse should be replaced at once.
Are the cables that lead from the consumer unit in good condition?	The cables should be fixed securely, with no bare wires showing. If the cables seem to be insulated with rubber (◁) have the whole of the insulation checked as soon as possible. Rubber insulation has a limited life, so yours could already be dangerous.
Is the earth connection from the consumer unit intact and in good condition (◁)?	If the connection seems loose or corroded have the Electricity Board check on whether the earthing is sound. You can check an RCCB (◁) by pushing the test button to see if it is working mechanically.
What is the condition of the fixed wiring between floors and in the loft or roof space?	If the cables are rubber-insulated have the system checked by a professional, but first examine each of the circuits, as they may not all have been renewed at the same time. If cable is run in conduit it can be hard to check on its condition, but if it looks doubtful where it enters accessories have the circuit checked professionally. Wiring should be fixed securely and sheathing should run into all accessories, with no bare wire in sight. Junction boxes on lighting circuits should be screwed firmly to the structure and should have their covers in place.
Is the wiring discreet and orderly?	Tidy all surface-run wiring into straight properly-clipped runs. Better still, bury the cable in the wall plaster or run it under floors and inside hollow walls.
Are there any old round-pin socket outlets?	See that their wiring is adequate, though old radial circuits should be replaced with modern ring circuits and 13amp square-pin sockets.
Are the outer casings of all accessories in good condition and fixed securely to the structure?	Replace any cracked or broken components and secure any loose fittings.
Do switches on all accessories work smoothly and effectively?	Where not, replace the accessories.
Are all the wires inside accessories attached securely to their terminals?	Tighten all loose terminals and ensure that no bare wires are visible. Fit sleeves to earth wires where missing.

QUESTIONS	ANSWERS
Is the insulation round wires dry and crumbly inside any accessories?	If so it is rubber insulation in advanced decomposition. Replace the covers carefully and have a professional check the system as soon as possible.
Do any sockets, switches or plugs get warm when live? Is there a smell of burning, or scorch marks on sockets or round the bases of the pins of plugs? Do sockets spark when you remove a plug, or switches when you operate them?	These things mean loose connections in the accessory or plug, or a poor connection between plug and socket. Tighten loose connections and clean all fuse clips, fuse caps and the pins of plugs with fine wire wool. If the fault persists try a new plug; lastly replace the socket or switch.
Is it difficult to insert a plug in a socket?	The socket is worn and should be replaced.
Are your sockets in the right places?	Sockets should be placed conveniently round a room so that you need never have long flexes trailing across the floor or under carpets. Add sockets to the ring circuit by running spurs.
Do you have enough sockets?	If you have to use plug adaptors you need more sockets. Replace singles with doubles, add spurs or extend the ring circuit.
Is there old, braided, twin flex hanging from some ceiling roses?	Replace it with PVC-insulated flex (▷). Also check that the wiring inside the rose is PVC-insulated.
Are there earth wires inside your ceiling roses?	If not get professional advice on whether to replace the lighting circuits.
Is your lighting efficient (▷)?	Make sure you have two-way switching on stairs, and consider extra sockets or different light fittings to make the lighting more effective or atmospheric (▷).
Is there power in the garage or the workshop?	Outbuildings separate from the house need their own power supply (▷).

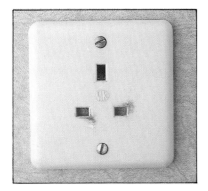

From left to right:

Scorch marks
Scorch marks on a socket or round the base of plug pins indicates poor connections.

Overloaded socket
If you have to use an adaptor to power your appliances fit extra sockets.

Unprotected connections
Make sure covers or faceplates are fitted to all accessories.

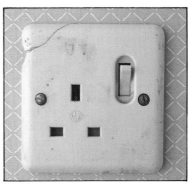

From left to right:

Incorrect fuse
Replace improper wire with fuse wire.

Round-pin socket
Replace a round-pin socket with a 13amp square-pin version.

Damaged socket
Replace cracked or broken faceplate.

POWER CIRCUITS: SURFACE-MOUNTING SOCKET OUTLETS

Whatever the type of circuits in your home, use only standard 13amp square-pin sockets. All round-pin sockets are now out of date, and though they may not be actually dangerous at the moment you should have them checked and consider changing the wiring to accommodate new 13amp socket outlets.

Before you start work on any socket switch the power off at the consumer unit and remove the fuse for that circuit, then test the socket with an appliance that you know to be working so as to be sure that the socket has been properly switched off.

Triple sockets
Triple sockets are useful in a situation where several electrical appliances are grouped together.

TYPES OF 13AMP SOCKET

Though all sockets are functionally very similar there are several variations on the basic component.

There are single and double sockets, and both are available either switched or unswitched and with or without neon indicators which tell you at a glance whether the socket is switched on. All these sockets are wired in the same way.

Another basic difference is in how sockets are mounted. They can be surface-mounted – screwed to the wall in a plastic box or pattress – or flush-mounted in a metal box buried in the wall with only its faceplate visible.

Switched single   **Unswitched single**

Switched double

Single switched with indicator

Positioning sockets

Decide on the most convenient positions for television, hi-fi, table lamps and so on and position sockets accordingly. To avoid using adaptors or long leads distribute the sockets evenly round the living room and bedrooms, and wherever possible fit doubles rather than singles. Don't forget sockets for running the vacuum cleaner in hallways and on landings.

The optimum height for a socket is 225 to 300mm (9in to 1ft) above the floor. This will clear most skirting boards and leave ample room for flexible cord to hang from a plug, while being high enough to be in no danger of getting struck by the vacuum cleaner. In the kitchen fit at least four double sockets 150mm (6in) above the worktops, more if you have a lot of small appliances. In addition fit sockets for floor-standing appliances like the refrigerator and dishwasher.

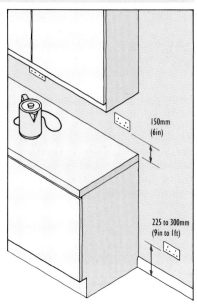

Optimum heights for socket outlets

Fixing to masonry

First break out the thin plastic webs that cover the fixing holes in the back of the pattress. The best tool for this is an electrician's screwdriver. Two fixings should be sufficient. The fixing holes are slotted to allow for adjustment.

Hold the pattress against the wall, levelling it at the same time with a small spirit level, and mark the fixing holes on the wall with a bradawl through the holes in the pattress. Drill and plug the holes with No 8 wall plugs.

With a larger screwdriver and pliers break out the plastic web covering the most convenient cable-entry hole in the pattress. For surface-run cable this will be in the side; for buried cable it will be the one in the base.

Feed the cable into the pattress to form a loop about 75mm (3in) long **(1)**, then fix the box to the wall with 32mm (1¼in) countersunk woodscrews.

Finally wire and fit the socket (◁).

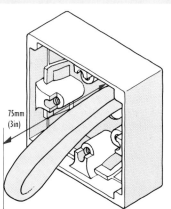

1 Leave a 75mm (3in) loop of cable at the box

Fixing to a hollow wall

On a dry partition or lath-and-plaster wall a surface-mounted pattress is fixed with any of the standard fixings for use on hollow walls (◁), though you can use ordinary woodscrews if you can position the pattress over a stud. In the latter case be sure you can feed the cable into it past the stud **(2)**.

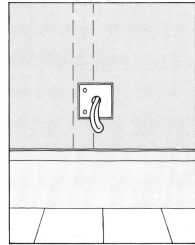

2 Feed the cable into the box past the stud

FLUSH-MOUNTING SOCKET OUTLETS

Fixing to masonry

Hold the metal box against the wall and draw round it with a pencil (1), then mark a 'chase' or channel running up from the skirting to the box's outline.

With a bolster and cold chisel cut away the plaster, down to the brickwork (2), within the marked area.

With a masonry drill bore several rows of holes down to the required depth (3) across the recess for the box, then with a cold chisel cut away the brick to the depth of the holes so that the box will lie flush with the plaster.

Try the box in the recess. If it fits in snugly mark the wall through the fixing holes in its back, then drill the wall for the screw plugs. If you have made the recess too deep, or the box rocks from side to side, apply some filler in the recess and press the box into it, flush with the wall and properly positioned. After about 10 minutes ease the box out carefully and leave the filler to harden so that you can mark, drill and plug the fixing holes through it.

Knock out one or more of the blanked-off holes in the box to accommodate the cable. Fit a grommet into each hole to protect the cable's sheathing from the metal edges (4), feed the cable into the box and screw the box to the wall.

Plaster up to the box and over the cable chased into the wall, and when the plaster has hardened wire and fit the socket (▷).

Fixing to plasterboard

To fit a flush socket to a wall made of plasterboard over wooden studs trace the outline of the box in position on the wall, then drill a hole in each corner of the shape with a brace and bit and cut out the waste with a padsaw.

Punch out the blanked-off entry holes in the box, fit rubber grommets and feed the cable into the box.

Clip dry-wall fixing flanges to the sides of the box (5). These will hold it in place by gripping the wall from inside. Ease one side of the box, with flange, into the recess and then, holding the screw fixing lugs so as not to lose the box, manoeuvre it until both flanges are behind the plasterboard and the box sits snugly in the hole.

Now wire and fit the socket (▷). As you tighten the fixing screws the plasterboard will be gripped between the flanges and the faceplate.

1 Draw round the mounting box

2 Chop away the plaster with a cold chisel

3 Drill out the brickwork with a masonry bit

4 Fit a soft grommet in the cable-entry hole

GROMMET

5 Dry-wall fixing flanges clipped to a box

FLUSH MOUNTING TO LATH AND PLASTER

If you wish to fit a flush socket in a lath-and-plaster wall try to locate it over a stud or nogging (▷).

Mark the position of the box, cut out the plaster and saw away the laths with a padsaw. Try the box for fit and, if necessary, chop a notch in the woodwork until the box lies flush with the wall surface (1). Feed in the cable and screw the box to the stud before wiring and fitting the socket (▷).

If you cannot position the socket on a stud, cut away enough plaster and laths to make a slot in the wall running from one stud to the next. Between the studs screw or skew-nail a softwood nogging to which you can fix the box. Set the batten back from the front edges of the studs if that is necessary to make the box lie flush with the wall surface (2). Feed the cable into the box and make good the surrounding plaster before you wire and fit the socket.

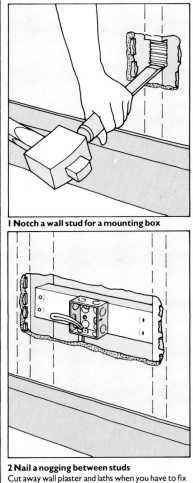

1 Notch a wall stud for a mounting box

2 Nail a nogging between studs
Cut away wall plaster and laths when you have to fix a mounting box between wall studs.

REPLACING SOCKET OUTLETS

If you have a socket outlet that needs to be replaced because it is faulty or *broken you should consider some options before undertaking the job.*

Simple replacement

Replacing a damaged socket with a similar one is quite straightforward. Any style will fit a flush-mounted box, but look carefully when you substitute a socket that screws to a surface-mounted pattress. Though it will fit and function perfectly well, square corners and edges on either will not suit rounded ones on the other. In such a case you may also have to buy a new, matching pattress.

An unswitched socket can be replaced with a switched one without any change to the wiring or fixing.

Switch off the power supply to the circuit, then remove the fixing screws holding the faceplate and pull the socket out of the box.

Loosen off the terminals and free the conductors. Check that all is well inside the pattress, then connect the conductors to the terminals of the new socket. Fit the faceplate, using the original screws if those supplied with the new socket don't match the thread in the pattress.

Surface to flush mounted

If you have to renew a socket for some reason you can use the occasion to replace a surface-mounted pattress with a flush box.

Turn off the power, remove the old socket and recess the metal box into the wall (◁), taking care not to damage the fixed wiring in the wall.

Replacing a single socket with a double

One way to increase the number of sockets in a room is to substitute doubles for singles. Any single socket on a ring circuit can be replaced with a double with no change to the wiring. You can similarly replace a single socket on a spur. Consider using the safer, switched sockets. The wiring is identical.

Surface to surface
Replacing a surface-mounted single unit with a surface-mounted double is quite easy. Having removed the old socket, simply fix the new, double pattress to the wall in the same place.

Flush to surface
Though flush-mounted sockets are neater, you may not want the disturbance to decor of installing a double one. Instead you can fit a double surface-mounted socket over the buried box of the single one (1). Turn off the power and remove the socket, leaving the metal box and the wiring in place. Knock out the cable-entry hole in the plastic double pattress and feed the

cable through it. When the pattress is centred over the old box (◁) two fixing holes will line up with the fixing lugs on the buried box. Break out the plastic webs and fix the new pattress to the lugs with the screws that held the old socket in place. Wire up the new double socket and fit it.

Flush to flush
Switch off the power to the circuit, remove the old socket at its metal box, then try the new double box over the hole. You can centre the box over the hole or align it with one end (2), whichever is the more convenient. Trace the outline of the box on the wall and cut out the brickwork (◁).

To substitute a double socket in a hollow wall use a similar procedure, installing the socket by whichever method is most convenient (◁).

Surface to flush
To replace a single surface-mounted socket with a flush double one proceed as described above.

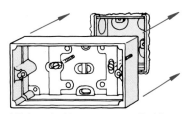

1 Fixing a double pattress over a flush box

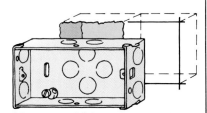

2 Cut out extra brickwork for a double box

CONNECTING UP TO A SOCKET

When a single cable is involved strip off the sheathing in the ordinary way and connect the wires to the terminals: the black wire to neutral – N, the red one to live – L and the earth wire, which you should insulate yourself with a sleeve, to earth – E (1). If necessary fold the stripped ends over so that no bare wire protrudes from a terminal.

When connecting to a ring circuit you can cut through the loop of cable, strip the sheathing from each half and twist together the bared ends of matching wires – live with live and so on – after slipping sleeves on the earth wires (2). Alternatively you can slit the sheathing lengthwise and peel it off, leaving the wires unbroken (3), bare a part of each wire by cutting away insulation, then pinch the exposed part of each wire into a tight fold with the pliers so that it will fit into its terminal. The second method ensures perfect contact, as the ring circuit is uninterrupted. You may have to cut the earth wire to slip sleeves over the halves.

Cable is stiff, and can make it hard to close the socket faceplate, so bend each conductor so that it will fold into the box or pattress. Locate both fixing screws and tighten them gradually in turns until the plate fits firmly in place against the wall or pattress.

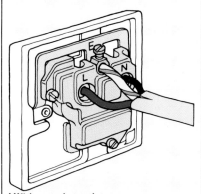

1 Wiring a socket outlet

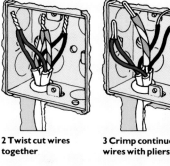

2 Twist cut wires together

3 Crimp continuous wires with pliers

ADDING A SPUR TO A RING CIRCUIT

If you need more sockets in convenient positions round a room you can run 2.5mm² spur cables from a ring circuit and have as many spurs as there are sockets already on the ring, each spur feeding a single or double socket.

A spur cable can be connected to any socket – or fused connection unit – on the ring, or to a junction box inserted in the ring, whichever is the easier. If running cable from one of the present sockets would mean disturbing the plaster it is more convenient to use a junction box, and if there is no socket within easy reach of the proposed new one using a junction box will save cable.

If cable is surface-run and you want to extend a row of sockets – behind a workbench, for example – it is simpler to connect the spur to a socket.

Examine the socket. If there is one cable feeding it, it is already on a spur; if there are three cables in the socket it is already feeding a spur itself. In either case you cannot connect a new spur, so look for a socket with two cables.

Connecting to an existing socket

Fix the new socket, wire it up in the ordinary way (See opposite) and run its spur cable to the existing socket. Switch off power and remove the existing socket. You may have to enlarge the entry hole in the pattress or knock out another to take the spur cable. Feed the cable into the pattress, prepare the conductors and twist their bared ends together with those of the matching conductors of the ring circuit. Insert the wires in their terminals: red – L, black – N and green/yellow – E and replace the socket. Switch the power back on and test the new spur socket.

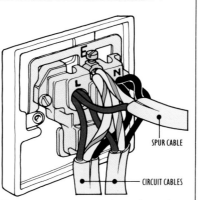

SPUR CABLE

CIRCUIT CABLES

Taking a spur from an existing socket outlet

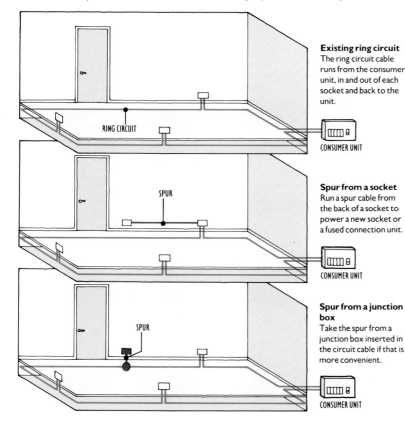

RING CIRCUIT

CONSUMER UNIT

Existing ring circuit
The ring circuit cable runs from the consumer unit, in and out of each socket and back to the unit.

SPUR

CONSUMER UNIT

Spur from a socket
Run a spur cable from the back of a socket to power a new socket or a fused connection unit.

SPUR

CONSUMER UNIT

Spur from a junction box
Take the spur from a junction box inserted in the circuit cable if that is more convenient.

CONNECTING TO A JUNCTION BOX

You will need a 30amp junction box with three terminals to connect to a ring circuit. It will have either knock-out cable-entry holes or a special cover that rotates to blank off unneeded holes. The cover must be screw-fixed.

Lift a floorboard close to the new socket and where you can connect to the ring circuit cable without having to stretch it.

Fix a platform for the box by nailing battens near the bottoms of two joists (See right) and screwing a 100 x 25mm (4 x 1in) strip of wood between the joists and resting on the battens. Loop the ring circuit cable over the platform before fixing it so that the cable need not be cut for connecting up. Remove the cover, screw the junction box to the platform and break out two cable-entry holes. If you do forget to loop the cable over the platform, just cut the cable when you come to connect it up.

Turn off the power at the consumer unit, then rest the ring circuit cable across the box and mark the amount of sheathing to remove. Slit it lengthwise and peel it off the conductors. Don't cut the live and neutral conductors, but slice away just enough insulation on each to expose a section of bare wire that will fit into the terminal (See right). Cut the earth wire and put insulating sleeves on the two ends.

Remove the screws from the terminals and lay the wires across them, the earth wire in the middle terminal and the live and neutral ones on the ends. Push the wires home with a screwdriver.

Having fitted and wired the new spur socket, run its cable to the junction box, cut and prepare the ends of the wires and break out an entry hole so that the spur wires can be fitted to the terminals of the box (See right). Take care that only colour-matched wires from both cables share terminals.

Replace the fixing screws, starting them by hand as they easily cross-thread, then tighten them with a screwdriver. Check that all the wires are secured and that the cables fit snugly in their entry holes with the sheathing running into the box, then fit the cover on the box.

Fix each cable to a nearby joist with cable clips, to take the strain off the terminals, then replace the floorboards.

Switch the power back on and test the new socket.

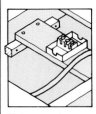

Make a wooden platform for a junction box

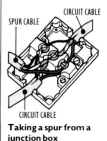

CIRCUIT CABLE

SPUR CABLE

CIRCUIT CABLE

Taking a spur from a junction box

EXTENDING A RING CIRCUIT

There are situations in which it is better to extend a ring circuit than to fit spurs. For instance, you may want to wire a room that was not adequately serviced before, or perhaps all of the conveniently placed sockets already have spurs running from them.

There are two ways of breaking into the ring: at an existing socket or via junction boxes. Whichever method you decide on, switch off the power to the circuit before you break into it.

Using an existing socket

Disconnect one in-going cable from a socket on the ring circuit and take this to the first new socket. Do it via a junction box if the cable will not reach otherwise. Continue the extension with a new section of cable from socket to socket, finally running it from the last one back to the socket where you broke into the ring. Joining the new cable to the old one within the socket completes the circuit.

Using junction boxes

Cut the ring cable and connect each cut end to a junction box, then run a new length of cable from one box to the other, looping it into the new sockets.

Running the extension

No matter how you plan to break into the ring always install the new work first and connect it up to the circuit only at the last moment. This allows you to use power tools on the extension. Switch the power off just before connecting up.

Decide on the positions of the new sockets and plan your cable run: an easy route is better than a difficult shorter one. Allow some slack in the cable.

Cut out the plaster and brickwork for sockets and cable and fit the boxes (◁). Now run the cable, leaving enough spare for joining to the ring circuit (◁), and take it up behind the skirting to the first socket. Leave a loop hanging from the box (See right), then take the cable on to the next, and so on until all the new sockets are supplied. Take the excess cable on to the point where you plan to join the ring.

Make good the plasterwork and fit the new sockets. Switch off the power, break into the ring and connect the extension to it. Switch the power on and test the new sockets separately.

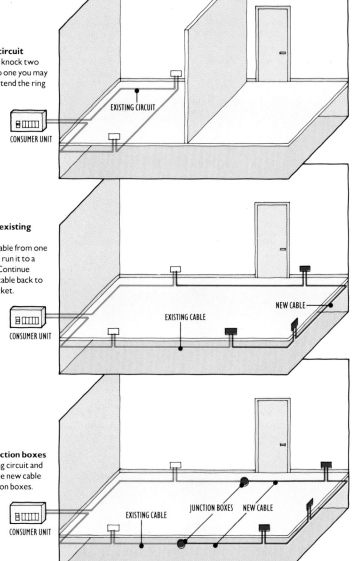

Existing circuit
When you knock two rooms into one you may need to extend the ring circuit.

EXISTING CIRCUIT

CONSUMER UNIT

Using an existing socket
Take the cable from one socket and run it to a new one. Continue with new cable back to the old socket.

NEW CABLE

EXISTING CABLE

CONSUMER UNIT

Using junction boxes
Cut the ring circuit and join it to the new cable with junction boxes.

EXISTING CABLE   JUNCTION BOXES   NEW CABLE

CONSUMER UNIT

LEAVE SOME SLACK IN THE CIRCUIT

Don't pull the cable too tight when you are running a new circuit: it puts a strain on the connections and makes it difficult to modify the circuit at a later stage should it become necessary.

Leave a generous loop of cable at each new socket position until you have run the complete circuit. At that stage you can pull the loop back ready for connecting to the socket.

Leave ample cable above the skirting

If, when you examine your installation, you find that the power circuit is radial you may decide to convert it to a ring circuit, particularly if you wish to supply a larger area (▷).

Checking cable and fuse

If the radial circuit is wired with 2.5mm^2 cable (solid conductors) continue the circuit back to the consumer unit with the same size cable but change the 20amp circuit fuse for a 30amp fuse and fuseway. If it is wired with 4mm^2 (stranded conductors) you can complete the ring with 2.5mm^2 cable and leave the 30amp circuit fuse alone.

The additional cable is run in exactly the same way as described for extending a ring circuit (See opposite). Join the new cable at the last socket on the radial circuit and run it to all the new sockets. From the last one run it to the consumer unit. Turn off the power.

Connecting to consumer unit

You should examine your consumer unit and familiarize yourself with it (▷). Even when the unit is switched off, the cable that connects the meter to the main switch is still live – so take care.

First locate the terminals to which the radial circuit is connected. The live (red wire) terminal is on the fuseway (or MCB) from which you removed the circuit fuse before starting the work. The neutral (black wire) terminal is on the neutral block, to which all of the black wires are connected. You can usually trace the black wire you're looking for by working along from the sheathed part of the cable, and the earth terminal similarly, by tracing the green-insulated conductor.

Pass the end of the new cable into the consumer unit as closely as possible to the original radial circuit cable. Cut it to length, strip off the sheathing and prepare the ends of the conductors.

Disconnect the live (red) conductor from its terminal and twist its end together with that of the red wire from the new cable, then reconnect both wires in the same terminal. Do the same for the black wires and then the green ones, but first slip a length of insulating sleeving over the new earth wire. Replace the cover.

Check that the circuit fuse is of the correct rating and then replace the fuse carrier. Close the consumer unit, switch on the power and test the circuit.

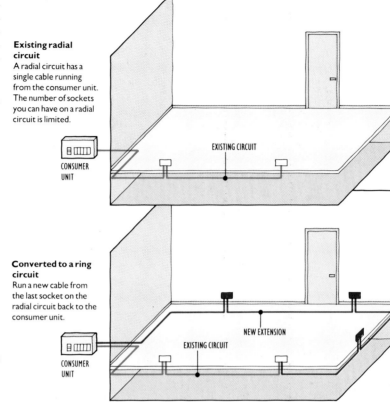

Existing radial circuit
A radial circuit has a single cable running from the consumer unit. The number of sockets you can have on a radial circuit is limited.

EXISTING CIRCUIT

CONSUMER UNIT

Converted to a ring circuit
Run a new cable from the last socket on the radial circuit back to the consumer unit.

NEW EXTENSION

EXISTING CIRCUIT

CONSUMER UNIT

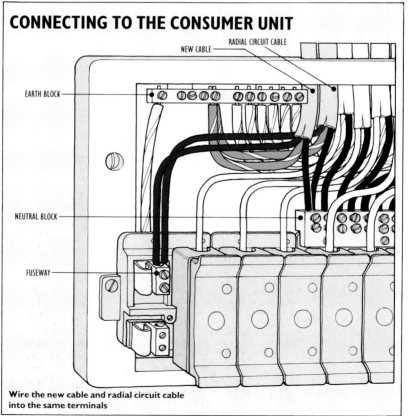

CONNECTING TO THE CONSUMER UNIT

NEW CABLE

RADIAL CIRCUIT CABLE

EARTH BLOCK

NEUTRAL BLOCK

FUSEWAY

Wire the new cable and radial circuit cable into the same terminals

FIXED APPLIANCES

A 13amp socket is designed to be flexible in use, enabling appliances to be moved from room to room and one socket to be used for different appliances at different times. But many appliances, large and small, are fixed to the structure of the house, or stand in one position permanently. Such appliances may just as well be wired permanently into the electrical installation. For some there is no alternative, and they may even require individual radial circuits direct from the consumer unit.

FUSED CONNECTION UNITS

A fused connection unit is basically a device for joining circuit wiring to the flex – or sometimes cable – of an appliance. The junction incorporates the added protection of a cartridge fuse like that found in a 13amp plug. If the appliance is connected by a flex, choose a unit with a cord outlet in the faceplate.

Some fused connection units are also switched, with or without a neon indicator that shows when the switch is on. The switched connection unit allows you completely to isolate the appliance from the mains.

All fused connection units are single – there are no double ones – with square faceplates that fit the standard plastic surface-mounted pattresses or the metal boxes for flush mounting.

Changing a fuse
With the power off, remove the retaining screw in the face of the fuse holder. Take the holder from the unit, prise out the old fuse and fit a new one. Replace the fuse holder.

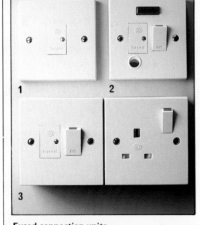

Fused connection units
1 Unswitched connection unit
2 Switched unit with cord outlet and indicator
3 Connection unit and socket in a dual box

Small appliances

All small electrical appliances with ratings of up to 3000W (3kW) – wall heaters, cooker hoods, heated towel rails and so on – can be wired into a ring or radial circuit by means of fused connection units. They could also be connected by 13amp plugs to sockets, but the electrical contact is not so good and there is some risk of fire with that type of permanent installation.

Always remember to switch off the power at the consumer unit before wiring a fused connection unit to the house circuitry.

Mounting a fused connection unit

A fused connection unit is mounted in the same type of box as an ordinary socket outlet, and the box is fixed to the wall in the same way (◁). The unit can also be mounted in a dual box which is designed to hold two single units – for example, a standard socket outlet next to a connection unit. The socket is wired to the ring circuit and the two units are linked together inside the box by a short 2.5mm² spur.

A dual mounting box

Wiring a fused connection unit

Fused connection units can be supplied by a ring circuit, a radial circuit or a spur.

Some appliances are connected to the unit by a length of flex while others are wired up with cable but the wiring arrangements inside the units are the same. Units with cord outlets have clamps to secure the connecting flex.

An unswitched connection unit has two live (L) terminals, one marked 'Load' for the brown wire of the flex, and the other marked 'Mains' for the red wire from the circuit cable. The blue wire from the flex and the black wire from the circuit cable go to similar neutral (N) terminals and both earth wires are connected to the E terminal or terminals (**1**).

A switched connection unit

A fused connection unit with a switch has two sets of terminals. Those marked 'Mains' are for the spur or ring cable that supplies the power; the terminals marked 'Load' are for the flex or cable from the appliance.

Wire up the flex side first, connecting the brown wire to the L terminal and the blue one to the N terminal, both on the Load side. Connect the green/yellow wire to the E terminal (**2**) and tighten the cord clamp.

Attach the circuit conductors to the Mains terminals – red to L and black to N, then sleeve the earth wire and take it to the E terminal (**2**).

If the fused connection unit is on a ring circuit you must fit two circuit conductors into each Mains terminal and the earth terminal.

Before securing the unit in its box with the fixing screws make sure the wires are held firmly in the terminals and that they can fold away neatly.

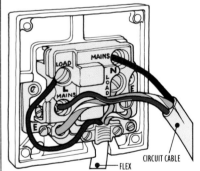

1 Wiring a fused connection unit

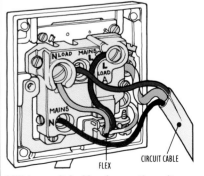

2 Wiring a switched fused connection unit

WIRING HEATERS

Skirting heaters, wall-mounted heaters and oil-filled radiators should be wired to a fused connection unit mounted nearby, between 150mm (6in) and 300mm (1ft) from the floor. Whether the connection to the unit is by flex or cable will depend on the type of heater. Follow the manufacturer's instructions, and fit a 13amp cartridge fuse in the connection unit.

In a bathroom a fused connection unit must be mounted out of reach, so any heater mounted near the floor of a bathroom must be on a cable to a connection unit installed outside the room. If the appliance is fitted with flex, mount a flexible cord outlet (1) next to the appliance, run a cable from the outlet to the fused connection unit outside and connect it to the 'Load' terminals in the unit.

The flex outlet is mounted on a standard surface-mounted pattress or flush on a metal box. At the back of the faceplate are three pairs of terminals to take the conductors from the flex and the cable (2).

Radiant wall heaters for use in bathrooms must be fixed high on the wall and out of reach from shower or bath. A fused connection unit, fitted with a 13amp fuse (or a 5amp fuse for a heater of 1kW or less), must be mounted at the same level and the heater must be controlled by a pull-cord double-pole switch – the type that works by breaking both live and neutral contacts. Many heaters have built-in double-pole switches; otherwise you must fit a ceiling-mounted 15amp double-pole switch between the connection unit and the heater (▷). Switch terminals marked 'Mains' are for the cable on the circuit side of the switch; those marked 'Load' are for the heater side. The earth wires are connected to a common terminal on the switch pattress.

If it is not possible to run a spur to the fused connection unit from a socket outside the bathroom don't be tempted to connect a radiant wall heater to the lighting circuit. Instead run a separate radial circuit from the fused connection unit to a 15amp fuseway in the consumer unit, using 2.5mm² cable.

Heated towel rail

The Regulations covering other kinds of heater apply to a heated towel rail if it is situated in a bathroom. As it is mounted near the floor, run a flex from it to a flexible-cord outlet which is in turn wired to a fused connection unit outside the bathroom.

Fit a 13amp fuse, or, for a heater of 1kW or less, a 5amp fuse.

If a heated towel rail is installed in a bedroom the fused connection unit can be mounted alongside it.

Heat/light unit

Heat/light units, which are usually fitted in bathrooms, incorporate a radiant heater and a light fitting in the one appliance. They are ceiling-mounted, usually in the position of the ceiling rose, but these units should never be connected to lighting circuits.

To install a heat/light unit in this position turn off the power and, having identified the lighting cables (▷), remove the rose and withdraw the cables into the ceiling void. Fit a junction box to a nearby joist and terminate the lighting cables at that point (3). Don't connect the switch cable, as it won't be needed.

From an unswitched fused connection unit mounted outside the bathroom run a 1.5mm² two-core and earth spur cable to a ceiling-mounted 15amp double-pole switch, and from there to the heat/light unit. Connect up to the fused connection unit (See opposite), then wire the heat/light unit according to the maker's instructions and fit a 13amp fuse in the connection unit.

SEE ALSO	
Details for: ▷	
Double-pole ceiling switch	43
Ceiling-rose connections	44
Switching off	16
Running cable	23-25

1 Flexible cord outlet

2 Wiring a flexible cord outlet

DISCONNECTED SWITCH CABLE

CIRCUIT CABLE

LIVE

EARTH

NEUTRAL

CIRCUIT CABLE

3 Terminating the lighting cables
Join the circuit cables in a junction box. Label the switch wire for future reference.

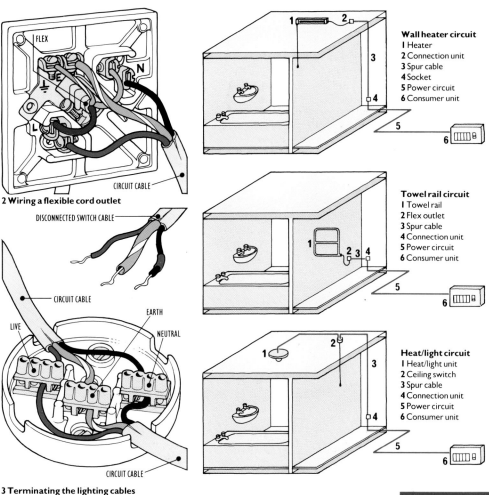

Wall heater circuit
1 Heater
2 Connection unit
3 Spur cable
4 Socket
5 Power circuit
6 Consumer unit

Towel rail circuit
1 Towel rail
2 Flex outlet
3 Spur cable
4 Connection unit
5 Power circuit
6 Consumer unit

Heat/light circuit
1 Heat/light unit
2 Ceiling switch
3 Spur cable
4 Connection unit
5 Power circuit
6 Consumer unit

WIRING SMALL APPLIANCES

Extractor fan

To install an extractor fan in a kitchen mount a fused connection unit 150mm (6in) above the worktop and run a cable to the fan or to a flex outlet next to it. If the fan has no integral switch use a switched connection unit to control it. Fit a 3 or 5amp fuse as recommended by the maker.

If the fan's speed and direction are controllable it may have a separate control unit, in which case wire the connection unit to it following the manufacturer's instructions.

To fit an extractor fan in a bathroom mount the fused connection unit outside and run the cable to the fan or its flex outlet via a ceiling-mounted double-pole pull-switch.

Fridges, dishwashers and washing machines

There is no reason why you cannot plug an appliance like a fridge, dishwasher or washing machine into a standard socket outlet except that in modern kitchens such appliances fit snugly under worktops, and sockets mounted behind them are hard to reach.

To control an appliance conveniently first mount a switched fused connection unit 150mm (6in) above the worktop and connect it to the ring circuit, then run a spur, using 2.5mm² cable, from the connection unit to a socket outlet mounted behind the appliance.

Cooker hood

Mount a fused connection unit, using a 3amp fuse, close to the cooker hood itself, or mount the connection unit at worktop height, then run a 1mm² cable from the unit to a flex outlet beside the cooker hood.

Instantaneous water heater

You can install an instantaneous water heater above a sink or washbasin to provide on-the-spot hot water. Join a 3kW model by heat-resistant flex to a switched fused connection unit mounted out of reach of anyone using the water.

If the heater is used in a bathroom wire it via a flex outlet to a ceiling pull-switch, then to the connection unit outside the bathroom. Lastly fit a 13amp fuse in the connection unit.

Wire a 7kW water heater like a shower (◁). If it is in the kitchen you can use a double-pole wall switch to control it.

Waste disposal unit

The waste disposal unit is housed in the sink base unit. Mount a switched fused connection unit 150mm (6in) above a worktop near the sink but out of reach of those using it and of small children.

From the unit run a 1mm² cable to a flex outlet next to the disposal unit.

Clearly label the connection unit 'disposal' to avoid accidents.

Finally fit a 3amp fuse.

Wall-mounted fan
Run a 1.5mm² cable from a fused connection unit to a wall-mounted extractor fan.

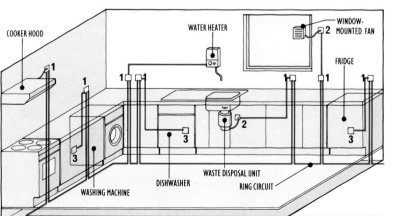

SHAVER SOCKETS

Special shaver outlets are the only sockets allowed in bathrooms. They contain transformers which isolate the user side of the units from the mains, so they cannot cause an electric shock.

A shaver unit can be wired to a spur from a ring circuit or to a junction box on an earthed lighting circuit. Connect the conductors to the shaver unit: red to L, black to N and earth to E **(1)**.

This type of socket conforms to the exacting British Standard, BS 3052, but there are shaver socket outlets which do not have isolating transformers. These are quite safe to install and use in a bedroom but must not be fitted in a bathroom. Wire such an outlet from the lighting circuit or from a fused connection unit on a ring circuit spur. Fit a 3amp fuse.

Shaver unit for use in a bathroom

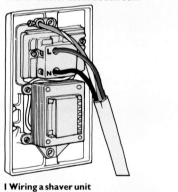

1 Wiring a shaver unit

Kitchen equipment circuits
1 Connection units
2 Flex outlets
3 Socket outlets

Powerful appliances such as cookers, with a power load greater than 3000W (3kW), must have their own radial circuits connected directly to the consumer unit, with separate fuses protecting them.

Cookers

Some small table cookers and separate ovens, which rate no more than 3000W (3kW), can consequently be connected to a ring circuit by a fused connection unit, or even by means of a 13amp plug and socket. But most domestic cookers are much more powerful, and must be installed on their own circuits.

The radial circuit

Cookers with a loading of up to 12000W (12kW) must be connected to a 30amp circuit. This requires a single radial circuit run in 6mm^2 two-core and earth cable protected by a 30amp circuit fuse.

For a cooker with a loading greater than 12000W you must install a 45amp circuit using 10mm^2 cable and a 45amp fuse. A separate radial circuit needs its own fuseway, and you can either use a spare fuseway in your consumer unit or fit an individual switchfuse unit, which will perform the same function as the consumer unit, but for a single appliance. Preferably choose a unit with a miniature circuit breaker; failing that, one with a cartridge fuse. As a last resort choose one that uses fuse wire.

Cooker control units

The cable from the consumer unit runs to a cooker control unit situated within 2m (6ft 6in) of the cooker.

The control unit is basically a double-pole isolating switch, but it may also incorporate a single 13amp switched socket outlet that can be used for an appliance such as an electric kettle. Now that more homes have a number of sockets installed at worktop height the additional one on the cooker control unit is not so important, and in fact it's better not to have one if it is to be situated above the cooker, from where a flex could trail across one of the hotplates.

The control unit must not only be within reach of the cooker but easily accessible, so don't install it inside a cupboard or under a worktop.

A single control unit can serve both parts of a split-level cooker, with separate cables running to the hob and the oven, provided the control unit is within 2m (6ft 6in) of both. If this is not possible in your case you will have to install a separate control unit for each part. The connecting cables must be of the same size as the cable used in the radial circuit.

Cooker control units can be surface- or flush-mounted.

Because a free-standing cooker has to be moved from time to time for cleaning round and behind it, it should be wired with enough cable to allow it to be moved well out from the wall. The cable is connected to a terminal outlet box which is screwed to the wall about 600mm (2ft) above floor level. A fixed cable runs from the outlet box to the cooker control unit.

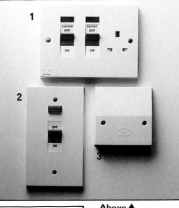

Above▲
1 Control unit with socket
2 Basic control unit
3 Terminal outlet box

Cooker circuit
1 Cooker
2 Terminal outlet box
3 Control unit
4 Radial circuit
5 Consumer unit

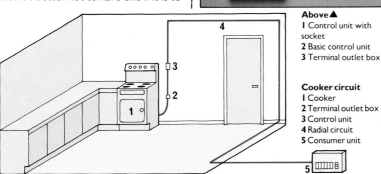

WIRING THE CONTROL UNIT

Having decided on the position for the cooker control unit knock out the cable-entry holes in the pattress and screw it to the wall. If your unit is to be flush-mounted cut a hole in the plaster and brickwork for the metal box (▷).

Running cable

Run and fix the cable, taking the most economical route to the cooker from the switchfuse unit or the consumer unit (▷). Cut a channel in the wall up to the control unit if you intend to bury the cable in the plaster, then cut similar channels for cables running to the separate hob and oven of a split-level cooker or for a single cable running to a terminal outlet box.

Connecting up the control unit

Feed the circuit cable and cooker cable into the control unit, then strip and prepare the conductors for connection. There are two sets of terminals in the control unit, one marked 'Mains' for the circuit conductors, and the other marked 'Load', for the cooker cable.

Run the red wires to the L terminals and the black ones to the terminals marked N. Put insulating sleeves on both earth conductors and connect them to the E terminals **(1)**. Fold the wires to fit into the box and screw on the faceplate.

SEE ALSO	
Details for: ▷	
Running cable	23-25
Flush mounting	29
Circuit fuses	19
Stripping cable	22
Switchfuse unit	38

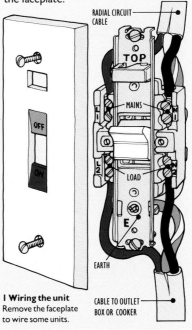

RADIAL CIRCUIT CABLE
TOP
MAINS
LOAD
E
EARTH
CABLE TO OUTLET BOX OR COOKER

1 Wiring the unit
Remove the faceplate to wire some units.

CONNECTING THE COOKER

Wiring to the cooker

Connect the cable to the hob and the oven following the manufacturer's instructions exactly.

For a free-standing cooker run the cable down the wall from the control unit to the terminal outlet box, which has terminals for connecting both cables. Strip and insert the wires of the control unit cable in the terminals **(1)**, then take the cooker cable and insert its wires in the same terminals, matching colour for colour, and secure it with the clamp. Screw the plastic faceplate onto the outlet box.

Wiring the switchfuse unit

If you are wiring to a fuseway in your consumer unit simply run the red wire to the terminal on the fuseway, the black one to the neutral terminal block and – having sleeved it – the earth wire to the earth terminal block (◁). All other connections will already be made. Don't forget to switch off the power before starting this work.

Here we will assume a cooker circuit to be run from a switchfuse unit. Screw the unit to the wall close to the consumer unit, feed the cooker circuit cable into it and prepare the conductors for connection. Fix the red wire to the live terminal on the fuse carrier or MCB, the black wire to the neutral terminal and the sleeved earth wire to the earth terminal **(2)**.

Prepare the meter leads, one black and one red, from PVC-sheathed and -insulated 16mm^2 single-core cable. (Use 10mm^2 cable if 16mm^2 cable is too thick for the switchfuse unit terminals but keep the meter leads as short as possible.)

Bare about 25mm (1in) of each cable and connect them to their separate terminals on the main isolation switch, red to L and black to N **(2)**. For an earth lead prepare a similar length of 16mm^2 single-core cable insulated in green/yellow PVC and attach it to the earth terminal in the switchfuse unit **(2)**. Fit the appropriate fuse, then plug in the fuse carrier. Label the carrier to indicate what circuit is run from the unit and fit the cover.

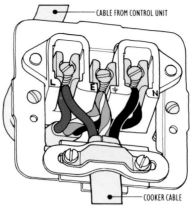

1 Wiring a terminal outlet box

2 Wiring switchfuse unit for the cooker

CONNECTING TO THE MAINS

A new circuit must be tested by a competent electrician and a certificate stating that the wiring complies with the Wiring Regulations must be submitted to the Electricity Board to apply for connecting to the mains. Do not attempt to make this connection yourself.

It may not be possible to attach both sets of meter leads – from consumer unit and switchfuse unit – to the meter, and the Board will have to install a service connector box with enough terminals to accommodate all the conductors. It's as well to consult the Electricity Board on these matters before starting.

IMMERSION HEATERS

Water in a storage cylinder is heated by an electric immersion heater, providing a central supply of hot water for the whole house. The heating element, rather like a larger version of the one that heats an electric kettle, is normally sheathed in copper, but more expensive sheathings of incoloy or titanium will increase the life of an element in hard water areas.

Adjusting the water temperature

A thermostat to control the maximum temperature of the water is set by adjusting a screw inside the plastic cap that covers the terminal box **(1)**.

Types of immersion heater

An immersion heater can be installed from the top of the cylinder or from the side, and top-entry units can have single or double elements. In the single-element top-entry type of heater the element extends down almost to the bottom of the cylinder, so that the whole of its contents is heated whenever the heater is switched on **(2)**. For economy one element in the double-element type is a short one that heats only the top half of the cylinder while the other element is a full-length one that is switched on when greater quantities of hot water are needed **(3)**.

A double-element heater with a single thermostat is called a twin-element heater. One with a thermostat for each element is known as a dual-element heater.

Side-entry heaters are the same length, one being positioned near the bottom of the cylinder and the other a little above half way **(4)**. This is a more efficient arrangement for heating water and controlling its temperature.

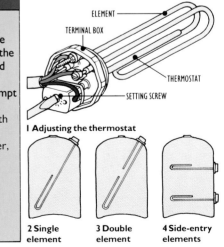

1 Adjusting the thermostat

2 Single element

3 Double element

4 Side-entry elements

HEATING WATER ON THE NIGHT RATE

If you agree to have a special meter installed the Electricity Board will supply you with power at a cheap rate for seven hours between midnight and 8.00 a.m., the hours varying with the time of year.

The scheme is called Economy 7 (▷). Providing you have a cylinder of big enough capacity to store hot water for a day's requirements you can benefit by producing all your hot water during the Economy 7 hours. Even if you heat your water electrically only in the summer it can be worth considering the scheme. For the water to retain its heat all day you must have an efficient insulating jacket fitted to the cylinder or a cylinder already factory-insulated with a layer of heat-retaining foam (▷).

If your cylinder is already fitted with an immersion heater you can use its wiring by fitting an Economy 7 programmer, a device which will switch your immersion heater on automatically at night and heat up the whole cylinder. If you should occasionally run out of hot water during the day you can adjust the programmer's controls to boost the temperature briefly on the more expensive daytime rate.

You can make even greater savings if you have two side-entry immersion heaters or a dual-element one. The programmer will switch on the longer element – or the bottom one – at night, but should you need daytime water-heating only the upper element is used.

You can have a similar arrangement without a programmer by wiring two separate circuits for the elements. The upper element is wired to the daytime supply and the lower one is wired to its own switchfuse unit and operated by the Electricity Board's Economy 7 timeswitch during the hours of the night-time tariff only. A setting of 75°C (167°F) is recommended for the lower element and 60°C (140°F) for the upper one. If you live in a soft water area or have heater elements sheathed in incoloy or titanium you can raise the temperatures to 80°C (175°F) and 65°C (150°F) respectively without reducing lives of the elements.

To ensure that you never run short of hot water leave the upper unit switched on permanently. It will start heating up only when the thermostat detects a temperature of 60°C (140°F), which should happen only rarely if you have a large and properly insulated cylinder.

WIRING THE IMMERSION HEATER

The circuit

Most immersion heaters are rated at 3kW, but while you can usually wire a 3kW appliance to a ring circuit an immersion heater is seen as taking a continuous 3kW, even though rarely switched on continuously. A continuous 3kW load would seriously reduce a ring circuit's capacity, so immersion heaters must have their own radial circuits.

The circuit is run in 2.5mm² two-core and earth cable protected by a 15amp fuse, though a 20amp fuse can be fitted quite safely. Each element must have a two-pole isolating switch mounted near the cylinder, probably marked 'water heater' and having a neon indicator (1). From a flex outlet at the switch a 2.5mm² heat-resistant flex runs to the immersion heater.

If the cylinder is in a bathroom the switch must be inaccessible to anyone using the bath or shower. If this precludes a normal water-heater switch use a 20amp ceiling-mounted pull-switch with a mechanical ON/OFF indicator.

Wiring two side-entry heaters

For simplicity use two switches, one for each heater and marked accordingly.

Wiring the switches

Fix the mounting boxes to the wall, feed a circuit cable to each and wire them in the same way. Strip and prepare the wires, connect them to the 'Mains' terminals – red to L, black to N – sleeve the earth wire and fix it to the common earth terminal (2). Prepare a heat-resistant flex for each switch. At each take the green/yellow earth wire to the common earth terminal, the other wires to the 'Load' terminals – brown to L and blue to N (2) – tighten the flex clamps and screw on the faceplates.

Wiring the heaters

The flex from the upper switch goes to the top heater and that from the lower switch to the bottom one. At each one feed the flex through the hole in the cap and prepare the wires. Connect the brown wire to one terminal on the thermostat (the other terminal on the thermostat is already connected to the wire running to the L terminal of the heating element). Connect the blue wire to the N terminal and green/yellow wire to the E terminal (3) and replace the caps on the terminal boxes.

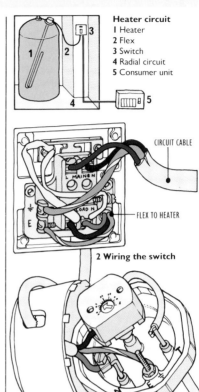

Heater circuit
1 Heater
2 Flex
3 Switch
4 Radial circuit
5 Consumer unit

CIRCUIT CABLE

FLEX TO HEATER

2 Wiring the switch

3 Wiring the heater

Running the cable

Run the circuit cables from the cylinder cupboard to the fuseboard and, with the power off, connect the cable from the upper heater to a spare fuseway in the consumer unit. Though the consumer unit is switched off the cable between main switch and meter is live, so take care. Wire the other cable to its own switchfuse unit – or storage-radiator consumer unit if you have one – ready for connecting to the Economy 7 timeswitch. Make the connections as described for a cooker circuit. (▷).

WIRING A DUAL-ELEMENT HEATER

Wire the circuit as described above but feed the flex from both switches into the cap on the terminal box of the heater. Strip and prepare the wires and connect both blue ones to the same N terminal and both earth wires to the E terminal. Connect the brown wire from the upper switch to the L1 terminal of the elements and the other brown wire to the L2 terminal (4).

SEE ALSO

Details for: ▷	
Insulation	61
Economy 7	6
Cooker circuit	37-38
Switching off	16
Consumer unit	18, 38

1 20amp switch for immersion heater

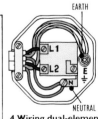

EARTH

L1

L2

E

N

NEUTRAL

4 Wiring dual-element

STORAGE HEATERS

The heart of a storage heater is a heat-retaining core, or block, that houses heating elements which are supplied with electricity during the night-time off-peak hours to take advantage of a cheap rate for power. The storage core is insulated in such a way that it will give off heat gradually during the day. Heat-emission is controlled in various ways.

With the earliest storage heaters it was not possible to control the rate of heat-emission, and towards the end of the day emission tended to lessen. This is no longer a problem. Modern heaters have dampers to regulate the flow of air through the core and control the rate of heat-loss. Some heaters have dampers that are automatically controlled by circuits that monitor the air temperature in the room.

Research has shown that a cold day is preceded by a proportionally cold night, and the more sophisticated storage heaters use this fact to store just the right amount of heat during the night to meet the needs of the following day.

Fan-assisted storage heaters have a similar heat-retaining core, efficiently insulated to reduce heat-loss to an absolute minimum. When the fan is switched on it draws air into the unit to be warmed before flowing out into the room. Apart from a very small amount of radiant heat through the casing, heat-emission occurs only when it is required, particularly if the fan is controlled thermostatically.

Storage heaters vary in size, with ratings from 1.2kW to 3.4kW. The fan-assisted ones are rated even higher, with loadings of up to 6kW. A large area needs a heater with a big heat-retaining core, able to store enough heat to warm it. As the lower-cost electricity is supplied for only a few hours a large core will need more powerful elements to charge it completely. The rate of heat-emission is not affected by the rating of a heater.

When you install storage heaters you have to assemble them yourself. Follow the maker's instructions and handle the heating elements and insulation carefully. Make sure that slim heaters are fixed securely to the walls, but with a 75mm (3in) gap all round so that air can circulate. Use fibre wall plugs for the fixings, as plastic ones can soften with heat.

Drying clothes on a storage heater will make a fusible link in the unit melt. Don't assemble old secondhand heaters as they may contain asbestos.

Storage heater circuits

Unlike other kinds of electrical heating the storage heaters in a house are all switched on at the same time, a procedure that would overload a ring circuit, so you have to provide an individual radial circuit for each heater. A separate consumer unit is installed to cope with off-peak load.

It's wise to choose a unit that is not only large enough to take all of the heater circuits but also has spare fuseways for possible additional heaters in the future. Make sure that there is an extra fuseway to take the immersion heater circuit so that your water can be heated at the night-time rate.

The circuit for storage heaters up to 4.8kW should be wired with 2.5mm² two-core and earth cable with a 20amp circuit fuse.

Outlets for storage heaters

The circuit cable for an ordinary storage heater should terminate at a 20amp double-pole switch with a flex outlet (1) that fits into a standard plastic or metal mounting box. A three-core heat-resistant flex connects the switch to the storage heater.

A fan-assisted heater needs a more complex circuit. The heating elements are supplied from a straightforward radial circuit using 2.5mm² cable, but the fan needs its own circuit for daytime use. Take a spur from a ring circuit to a fused connection unit that has a 3amp fuse (◁) and run a 1.5mm² two-core and earth cable from the unit for the fan. The heater and fan circuits both terminate at a special dual switch (2) where fan and heater can be isolated simultaneously. Two lengths of heat-resistant flex run from the switch, one to the heater, the other to the fan. The dual switch can be surface- or flush-mounted.

Electricity Board equipment

Because a storage heater system uses cheap-rate power to make it economical you need a special meter which calculates separately the number of units consumed during the night and those used in daytime. You also need a time switch to connect the various circuits at the appropriate time.

This equipment is supplied by the Electricity Board, which you should contact for advice as soon as possible if you plan to have storage heaters. At the same time make sure that your present electrical installation is safe, in particular the provision for earthing; otherwise the Board may be unwilling to connect the new circuits.

Storage heater circuits

1 Off-peak consumer unit
2 Day-time consumer unit
3 Radial circuits to heaters
4 20amp switch
5 Storage heater
6 Fan-assisted storage heater
7 Dual switch
8 Connection unit
9 Ring circuit

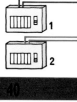

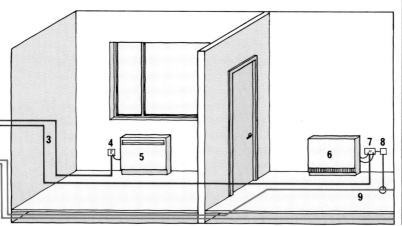

Wiring storage heaters

For ordinary storage heaters mount a 20amp switch close to where you will stand each heater. Run a single length of 2.5mm² two-core and earth cable from each switch to the site of the new consumer unit, taking the most economical route (▷).

Feed a cable into the mounting box of each switch, strip and prepare the wires and connect them to the 'Mains' terminals: red to L, black to N. Sleeve the earth wire and connect it to the E terminal (1).

Pass the flex from each heater through the outlet hole in the faceplate of its switch, strip and prepare the wires and connect them to the 'Load' terminals: brown to L, blue to N and the green/yellow earth wire to E (1). Tighten the cord clamp and fix the switch into its mounting box.

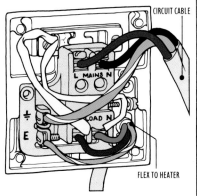

1 Wiring a 20amp switch for a storage heater

Wiring fan-assisted heaters

When you wire a fan-assisted heater mount a dual switch nearby, and from its 'heater' side (2) run a 2.5mm² two-core and earth cable to the consumer unit.

Mount a fused connection unit near the switch and run a short length of 1.5mm² two-core and earth cable between the two, connecting to the 'Load' side of the connection unit and the 'Fan' side of the dual switch (2).

Run a spur of 2.5mm² two-core and earth cable from the 'Mains' terminals on the connection unit (2) to a junction box or socket on the nearest ring circuit (▷).

Feed the fan and heater flex into the outlets in the faceplate of the dual switch and strip and prepare the wires. Connect each flex to its own part of the switch, which is clearly labelled (2).

Tighten the cord clamps and screw the switch to its box.

WIRING THE CONSUMER UNIT

When you buy a new consumer unit get one with miniature circuit breakers, or at least cartridge fuses. Install a 20amp fuse carrier or MCB for each heater circuit and a 15amp one for an immersion heater circuit if needed.

Mount the unit on an 18mm (¾in) plywood board, not on the Electricity Board's meter board even if there's room. Cut your board to size, knock out the entry holes in the back of the unit, lay it on the board and mark out the positions of the holes – for fuseways, meter leads and earth wire. Drill 18mm (¾in) holes in the board for the cables and paint or varnish both sides to damp-proof it.

Screw the board to the wall, using plastic or ceramic insulators to space it away so that damp won't penetrate it. Get the insulators when you buy the consumer unit. Position the board as close to the meter as you can so as to keep the meter leads short.

Screw the consumer unit to the board, run the circuit cables from the heaters into it, one at a time, and prepare the wires for connection.

The circuits are wired in the same way to separate fuseways: the red wire to the terminal on the fuseway, the black one to the neutral terminal block and the earth wire to the earth block after being sleeved green/yellow.

Use 16mm² single-core cable for the meter leads. They are insulated in red for the live conductor and black for the neutral. The outer sheathing may be in the same colours but it is often grey for both. Feed the leads into the consumer unit and connect them to their terminals – red to L, black to N – on the main isolating switch.

Connect a length of green/yellow 16mm² single-core cable to the earth block. Its other end will be connected by an Electricity Board representative to the consumer's earth terminal.

Fit MCBs or clip a fuse into each of the fuse carriers and insert the carriers into their fuseways. Label all of the circuits clearly so that in future you can tell which heater each one supplies.

Fit the covers on the consumer unit and arrange for the Electricity Board to test the circuits and connect the unit to the meter and earth. Don't, in any circumstances, try to make these connections yourself.

SEE ALSO

Details for: ▷	
Running cable	23-25
Running a spur	31
Stripping flex	13
Fuses/MCBs	19
Cables	22
Stripping cable	22

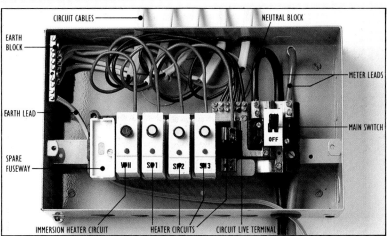

Wiring the consumer unit for storage heaters

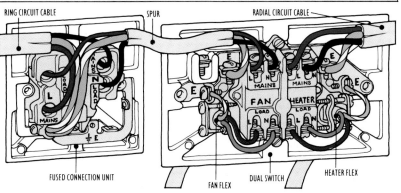

2 Wiring a dual switch
Connect a fused connection unit to the dual switch.

DOORBELLS, BUZZERS AND CHIMES

INSTALLING A SYSTEM

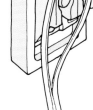

Chimes
A set of chimes has two tubes, each tuned to a different note.

I Wiring a bell push

Whether you choose a doorbell, a buzzer or a set of chimes there are no practical differences to affect the business of installing them.

Bells
Most doorbells are of the 'trembler' type. When electricity is supplied to the bell – that is when someone presses the button at the door – it activates an electro-magnet which causes a striker to hit the bell. But as the striker moves to the bell it breaks a contact, cutting off power to the magnet, so the striker swings back, makes contact again and repeats the process, going on for as long as the button is depressed. This type of bell can be operated by battery or by a mains transformer. Other types of doorbells, known as AC bells, can be used only with mains power.

Buzzers
A buzzer operates on exactly the same principle as a trembler bell but in the buzzer the striker hits the magnet itself instead of a bell.

Chimes
A set of ordinary door chimes has two tubes or bars tuned to different notes. Between them is a solenoid, a wound coil that acts like a magnet when it is energized. When the button is pressed a spring-loaded plunger inside the solenoid is thrown against one tube, sounding a note. When the button is released the spring throws the plunger against the other tube, sounding the other note before returning to its point of rest. Other chimes have a programmed microprocessor that gives a choice of tunes when operated by the bell push. Most chimes can be run from a battery or a transformer.

Bell pushes
When the bell push at the door is pressed it completes the circuit that supplies power to the bell. It is a switch that is on only when held in the 'on' position. Inside it are two contacts to which the circuit wires are connected. One contact is spring-loaded, touching the other when the push is depressed, to complete the circuit, and springing back when the push is released (**1**).

Illuminated bell pushes incorporate a tiny bulb which enables you to see the bell push in the dark. These must be operated from mains transformers, as the power to the bulb, though only a trickle, is on continuously and would soon drain a battery. Luminous types glow at night without a power supply.

Batteries or transformer?

Some doorbells and chimes house batteries inside their casings, while others incorporate built-in transformers that reduce the 240-volt mains electricity to the very low voltages needed for this type of equipment. For many bells or chimes you can use either method. Most of them use two or four 1½ volt batteries, but some require a 4½ volt battery, housed separately. Transformers sold for use with doorbell systems have three low-voltage tappings – 3 volt, 5 volt and 8 volt – to meet various needs. Usually 3 volt and 5 volt connections are suitable for bells or buzzers, and the 8 volt tapping is enough for many sets of chimes.

Some other chimes need higher voltages, and for these you will need a transformer with 4 volt, 8 volt and 12 volt tappings. A bell transformer must be designed so that the full mains voltage cannot cross over to the low-voltage wiring.

Circuit wiring

The battery, bell and push are connected by fine insulated 'bell wire', usually two-core. Being so fine, it is often surface-run, fixed to the skirting and door frame by small staples, but it can be run under floors and in cupboards. Bell wire also connects a transformer to a bell and bell push.

The transformer itself connects to a junction box or ceiling rose on a lighting circuit with 1mm² two-core and earth cable; it must be earthed, so if your lighting system has no earth wire use another method.

Run a spur from a ring circuit in 2.5mm² two-core and earth cable to an unswitched fused connection unit fitted with a 3amp fuse, then a 1mm² two-core and earth cable from the unit to the transformer 'Mains' terminals.

Alternatively, you can run 1mm² two-core and earth cable from a spare 5amp fuseway in a consumer unit to the transformer 'Mains' terminals.

The bell itself can be installed in any convenient position except over a source of heat. The entrance hall is usually best as a bell there can be heard in most parts of the house. Keep the bell wire runs to a minimum, especially for a battery-operated bell. With a mains-powered bell you will not want long and costly runs of cable, so place the transformer where it can be wired simply. A cupboard under the stairs is a good place, especially if it is near the consumer unit.

Drill a small hole in the door frame and pass the bell wire through to the outside. Fix the conductors to the terminals of the push, then screw it over the hole.

If the battery is in the bell casing there will be two terminals for attaching the other ends of the wires. Either wire can go to either terminal. If the battery is separate from the bell run the bell wire from the push to the bell. Separate the conductors, cut one of them and join each cut end to a bell terminal. Run the wire on to the battery and attach it to the terminals (**1**).

When you wire to a transformer proceed as above but connect the bell wire to whichever two of the three terminals combine to give you the required voltage (**2**). Some bells and chimes need separate lengths of bell wire, one from the bell push and another from the transformer. Fix the wires to terminals in the bell housing following manufacturers' instructions.

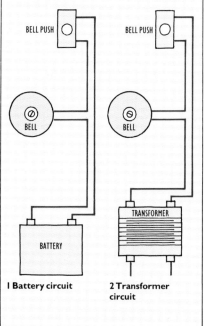

I Battery circuit

2 Transformer circuit

WIRING A SHOWER UNIT

An electrically heated shower unit is plumbed into the mains water supply (▷). The water pressure operates a switch to energize a heater that heats the water on its way to the shower head. Because there is so little time to heat the flowing water instantaneous showers use a heavy load, from 6 to 8kW. Consequently a shower needs a separate radial circuit.

The circuit cable must be 10mm² two-core and earth, protected by a 45amp fuse in a spare fuseway at the consumer unit or in a separate 45amp switchfuse unit. The cable runs directly to the shower unit where it must be wired according to the manufacturer's instructions.

The shower has its own on/off switch, but there must also be a separate isolating switch in the circuit. This must not be accessible to anyone using the shower, so install a ceiling-mounted 45amp double-pole pull-cord switch with a contact gap of at least 3mm, and preferably one with a neon 'on' indicator. Fix the backplate of the switch to the ceiling (▷) and, having sleeved the earth conductors, connect them to the E terminal on the switch. Connect the conductors from the consumer unit to the 'Mains' terminal of the switch and those of the cable to the shower to the 'Load' terminals (**1**).

Shower unit, metallic pipes and fittings must be bonded to earth (▷).

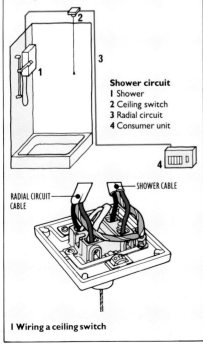

Shower circuit
1 Shower
2 Ceiling switch
3 Radial circuit
4 Consumer unit

RADIAL CIRCUIT CABLE

SHOWER CABLE

1 Wiring a ceiling switch

LIGHTING CIRCUITS

Every lighting system needs a feed cable to supply power to all the lighting points, and a switch that can interrupt the supply to each point. There are two ways of meeting these requirements in your home: the junction-box system and the loop-in system. Your house may be wired with either one, though it is quite likely that there will be a combination of the two systems.

The junction-box system
In the junction-box system a two-core and earth feed cable runs from a fuseway in the consumer unit to a series of junction boxes, one for each lighting point. From each junction box a separate cable runs to a light and another runs to its switch.

The loop-in system
In the loop-in system the ceiling rose takes the place of the junction box. The cable from the consumer unit runs into each rose and out again, then on to the next. The switch cable and the flex to the bulb are connected at the rose.

Combined system
The loop-in system is now the most widely used as it entails fewer connections as well as saving on the cost of junction boxes. However, lights at some distance from a loop-in circuit are often run from a junction box on the circuit to save cable, and lights added after the circuit has been installed are often wired from junction boxes.

The circuit
Both the junction-box and the loop-in systems are, in effect, multi-outlet radial circuits. The cable runs from the consumer unit, looping in and out of the ceiling roses or junction boxes and terminating at the last one. It doesn't return to the consumer unit like the cable of a ring circuit. Lighting circuits require 1mm² PVC-insulated and -sheathed two-core and earth cable. Each circuit is protected by a 5amp circuit fuse, and so up to twelve 100W bulbs or their equivalent can use the circuit. In the average house it is practical to have two separate lighting circuits, one for the ground floor and another for one upstairs.

Junction-box system
1 Consumer unit
2 Circuit cable
3 Junction box
4 Light cable
5 Switch cable

Loop-in system
1 Consumer unit
2 Circuit cable
3 Ceiling rose
4 Switch cable

● **Plumbing a shower**
Connect an instantaneous shower to the rising main with a 15mm (½in) supply pipe. Attach the pipe to the shower unit with a tap connector.

IDENTIFYING THE CONNECTIONS

Loop-in system

A modern loop-in ceiling rose has three terminal blocks arranged in a row. The live (red) conductors from the two cut ends of the circuit-feed cable run to the central live block, and the neutral (black) conductors run to the neutral block on one side. The earth conductors run to a common earth terminal (1).

The live (red) conductor from the switch cable is connected to the remaining terminal in the central live block. Power runs through this conductor to the switch and back to the ceiling rose through the black conductor, the 'switch-return wire', and this is connected to the third terminal block in the ceiling rose, the 'switch-wire block'. When the light is 'on' the switch-return wire is live, so it should be identified with a piece of red tape wrapped round it to distinguish it from the other black wires, which are neutral. The earth conductor in the switch cable goes to the common earth terminal (1).

The brown conductor from the pendant-light flex connects to the remaining terminal in the switch block while the blue conductor runs to the neutral block. If three-core flex is used the green/yellow earth conductor runs to the common earth terminal (1).

When the circuit-feed cable terminates at the last ceiling rose on the circuit only one set of cable conductors is connected (2). Switch cable and light flex are connected like those in a normal loop-in rose.

Junction-box system

The junction boxes on a lighting circuit normally have four unmarked terminals, for live, neutral, earth and switch connections. The live, neutral and earth conductors from the circuit feed cable go to their respective terminals (3).

The live conductor from the cable that runs to the ceiling rose is connected to the switch terminal, the black wire to the neutral terminal and the earth conductor to the earth terminal (3).

The red wire from the switch cable is connected to the live terminal, the earth conductor to the earth terminal and the black return wire from the switch goes to the switch terminal (3).

This last conductor should be identified by having a piece of red tape wrapped round it.

At the ceiling rose the live cable conductor is connected to one of the outer terminal blocks, the neutral conductor to the other and the central block left empty. The earth conductor goes to the earth terminal (4).

The flex conductors are wired up to match those from the cable. The brown wire is connected to the same terminal block as the red conductor and the blue wire goes to the block holding the black conductor. If the flex has a yellow/green earth wire it is connected to the common earth terminal (4).

Checking an old light circuit

Switch off the power at the consumer unit, remove the circuit fuse and examine ceiling roses and light switches for any signs of deterioration. Pre-World War II wiring will have been carried out in rubber-insulated and -sheathed cable. If the insulation seems dry and crumbly it is no longer safe. The circuit should be rewired. If you detect any signs at all that the circuitry is out of date, and perhaps dangerous, consult a professional electrician.

An old installation may have loop-in or junction-box lighting circuits, though the junction-box system is more likely. It may also lack any earth conductors, another good reason for renewing it.

Old fabric-covered flex should be replaced

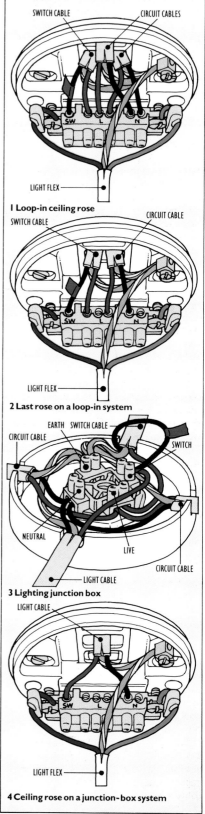

CONNECTIONS FOR LOOP-IN AND JUNCTION-BOX SYSTEMS

SWITCH CABLE   CIRCUIT CABLES

SW   L   N

LIGHT FLEX

1 Loop-in ceiling rose

SWITCH CABLE   CIRCUIT CABLE

SW   L   N

LIGHT FLEX

2 Last rose on a loop-in system

EARTH   SWITCH CABLE

CIRCUIT CABLE   SWITCH

NEUTRAL   LIVE

CIRCUIT CABLE

LIGHT CABLE

3 Lighting junction box

LIGHT CABLE

SW   L   N

LIGHT FLEX

4 Ceiling rose on a junction-box system

PLANNING YOUR LIGHTING

Successful lighting must be functional to enable you to work efficiently and read or study without eyestrain. It must define areas of potential danger and provide general background illumination. But the decorative element of lighting is equally important. It can create an atmosphere of warmth and wellbeing, highlight objects of beauty or interest and transform an interior with areas of light and shadow.

Living areas

For the living areas of the house, the accent should be one of versatility, creating areas of light where they are needed most, both for function and dramatic effect. Seating areas are best served by lighting placed at a low level so that naked bulbs are not directed straight into the eyes, and in such a position that the pages of a book or newspaper are illuminated from beside the reader. Choose lighting which is not so harsh that it would cause glare from white paper, and supplement it with additional low-powered lighting to reduce the effects of contrast between the page and darker areas beyond.

Working at a desk demands similar conditions but the light source must be situated in front of you to avoid throwing your own shadow across the work. Choose a properly shaded desk lamp, or conceal lighting under wall storage or bookshelves above the desk.

Similar concealed lighting is ideal for a wall-hung hi-fi system but you may require extra lighting in the form of ceiling-mounted downlighters to illuminate the shelves themselves. Alternatively, use fittings designed to clip onto the shelves or wall uprights.

Concealed lighting in other areas of the living room can be very attractive. Strip lights placed on top and at the back of high cupboards will bounce light off the ceiling into the room. Hide lighting behind pelmets to accentuate curtains, or along a wall to light pictures. Individual artworks can be picked out with specially designed strip lights placed above them, or use a ceiling spot light which is adjustable to place the pool of light exactly where it is required. Take care with pictures protected by glass as reflections will destroy the desired effect. Use lighting in an alcove or recess to give maximum impact to an attractive display of collected items.

SEE ALSO

Details for: ▷	
Lighting circuits	43-53
Light fittings	47

Concealed lighting
Interesting effects are created by concealing the actual source of artificial light and allowing it to bounce off adjacent surfaces to illuminate areas of the room.

Atmospheric lighting
Carefully placed light fittings produce an atmosphere of warmth and wellbeing.

PLANNING YOUR LIGHTING

Sleeping areas

Bedside lamps are essential requirements in any bedroom, or better still, use concealed lighting above the bedhead. Position the fitting low enough to prevent light falling on your face as you lie in bed. Install two lights behind the baffle over a double bed, each controlled individually so that your partner can sleep undisturbed if you want to read into the early hours. A dressing table needs its own light source placed so that it cannot be seen in the mirror but illuminates the person using it. Wall lights or downlighters in the ceiling will provide atmospheric lighting but install two-way switching (◁) so that you can control them from the bed and the door. Make sure bedside light fittings in a child's room are completely tamper-proof and, preferably, double-insulated (◁). A dimmer switch controlling the main room lighting will provide enough light to comfort a child at night but can be turned to full brightness when he or she is playing in the evening.

Dining areas/kitchens

A rise-and-fall unit is the ideal fitting to light a dining table because its height can be adjusted exactly. If you eat in the kitchen, have separate controls for the table lighting and work areas so that you can create a cosy dining area without having to illuminate the rest of the room. In addition to a good background light, illuminate kitchen worktops with strip lights placed under the wall cupboards but hidden from view by baffles along the front edges. Place a track light or downlighters over the sink to eliminate your own shadow.

Bathrooms

Safety must be your first priority when lighting a bathroom. Fittings must be designed to protect electrical connections from moisture and steam, and they must be controlled from outside the room or by a ceiling-mounted switch. It can be difficult to create atmospheric lighting in a bathroom, but concealed light directed onto the ceiling is one solution as long as you provide another source of light over the basin mirror.

Stairways

Light staircases from above so that treads are illuminated clearly, throwing the risers into shadow. This will define the steps clearly for anyone with poor eyesight. Place a light over each landing or turn of the staircase. Two- or even three-way switching is essential to be certain that no-one has to negotiate the stairs in darkness.

Workshops

Plan workshop lighting with efficiency and safety in mind. Light a workbench like a desk and provide individual, adjustable fittings for machine tools.

Bedside lamps
(Below)
You can expend a great deal of thought on planning the lighting in your bedroom only to find that two simple bedside lamps are the perfect solution.

Bathroom lighting
(Right)
Light fittings concealed behind a translucent screen produce an original and safe form of lighting for a bathroom.

Concentrating areas of illumination
(Above)
Spotlights will concentrate pools of light to illuminate the functional area of a kitchen and a dining table.

An improvised lampshade
(Left)
This Oriental sunshade will throw a diffused light on the dining table while bouncing extra light off the ceiling.

LIGHT FITTINGS

There is now a vast range of lighting fittings that can be used in the home, and though they may differ greatly in their appearance they can be grouped roughly in about eight basic categories according to their functions.

Types of light fitting

Pendant lights
The pendant light is probably the most common light fitting. It comprises a lamp-holder with bulb, usually with some kind of shade, suspended from a ceiling rose by a length of flex. The flex is connected to the power supply through terminals inside the ceiling rose (See opposite).

Decorative pendant lights
Most decorative pendant light fittings are designed to take several bulbs, and are consequently much heavier than standard pendant lights. Because of its weight this type of fitting is attached to the ceiling by a rigid tube. The flex that conducts the power to the bulbs passes through the tube to the lighting circuit.

Close-mounted ceiling lights
A close-mounted ceiling light is screwed directly to the ceiling, dispensing with a ceiling rose, by means of a backplate that houses the lampholder or holders. The fitting is usually enclosed by some kind of rigid light-diffuser that is also attached to the backplate.

Recessed ceiling lights
In this type of light fitting the lamp housing itself is recessed into the ceiling void and the diffuser lies flush with, or projects only slightly below, the ceiling. Lights of this type are ideal for rooms with low ceilings; they are often referred to as downlighters.

Track lights
Several individual light fittings can be attached to an aluminium track which is screwed to the ceiling or wall. Because a contact runs the length of the track, lights can be fitted anywhere along it.

Fluorescent light fittings
A fluorescent light fitting uses a glass tube containing mercury vapour. The power makes electrons flow between electrodes at the ends of the tube and bombard an internal coating, which fluoresces to produce the light. The fitting, which also contains a starter mechanism, is usually mounted directly on the ceiling, though as they produce very little heat fluorescent lights are used for under-cupboard lighting.

Wall lights
A light fitting adapted for screwing to a wall instead of a ceiling can be supplied from the lighting circuit in the ceiling void or from a spur off the ring circuit. Various kinds of close-mounted fittings or adjustable spotlights are the most popular wall lights.

Batten holders
A batten holder is a basic fitting with a lamp-holder mounted on a plate that fixes directly to wall or ceiling. Straight, angled and swivel versions are available. Batten holders are for use in areas – such as lofts or cellars – where appearance is not important.

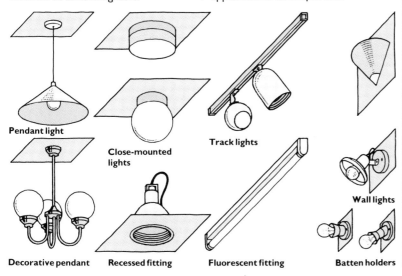

Pendant light

Close-mounted lights

Track lights

Decorative pendant

Recessed fitting

Fluorescent fitting

Wall lights

Batten holders

REPLACING A CEILING ROSE

Turn off power at the consumer unit and remove the circuit fuse. Switching off at the wall is not enough.

Unscrew the rose's cover and inspect the connections so that you can wire the new ones to work in the same way. A modern loop-in rose will be wired by one of the methods shown opposite. Identify the switch-return wire with tape if it is not already marked. If there is only one red and one black conductor the rose is on a junction-box system and will have no switch cable.

In an old rose you may have to identify the wires. If there are wires running into three terminal blocks look first for the one with all red wires and no flex wires. That is the live block, containing live circuit-feed wires and a live switch wire. The neutral terminal block contains the black neutral circuit feed wires and the blue flex wire. The third block will contain the brown flex wire plus a black conductor – the switch return wire – which should be marked with red tape, and may even be sheathed in red PVC.

All earth wires will run to one terminal on the backplate, but an old system may have no earth wires. In this case reconnect the other conductors temporarily but get expert advice on rewiring the circuit.

Fixing the new rose
Disconnect the wires from the terminals and separate any that are twisted together, but identify them with tapes. Unscrew the old backplate from the ceiling. Knock out the entry hole in the new backplate, thread the cables through it and fix the backplate to the ceiling, using the old screws and fixing points if possible. If the old fixings are not secure nail a piece of wood between the joists above the ceiling (See right) and drill a hole through it from below for cable access. Screw the new rose backplate to the wood through the ceiling.

Make sure that the ends of the conductors are clean and sound, then wire the ceiling rose, following the diagrams opposite.

Slip the new cover over the pendant flex and connect the flex wires to the terminals in the rose, looping the wires over the rose's support hooks to take the weight off the terminals. Screw the cover onto the backplate, switch on the power and test the light.

Fixing a platform
Scew-nail a board between the joists to support a ceiling rose.

FITTING A CLOSE-MOUNTED LIGHT

Some close-mounted light fittings have a backplate that screws directly to the ceiling in place of a ceiling rose. To fit one, first switch off the power for the circuit at the consumer unit and take out the fuse, then remove the ceiling rose and fix the backplate to the ceiling (◁).

1 BESA box
Use a BESA box to house the connections when a light fitting is supplied without a backplate.

If only one cable feeds the light attach its conductors to the terminals of the lampholder and the earth wire to the terminal on the backplate.

As more heat will be generated in an enclosed fitting, slip heat-resistant sleeving over the conductors before attaching them to their terminals.

If the original ceiling rose was wired into a loop-in system the light fitting will not accommodate all the cables. Withdraw them into the ceiling void and wire them into a junction box (◁) screwed to a length of 100 x 25mm (4 x 1in) timber nailed between the joists, then run a short length of heat-resistant cable from the junction box to the close-mounted light fitting.

FITTING A BESA BOX

Fix a wooden platform between the ceiling joists to support the junction box and the BESA box.

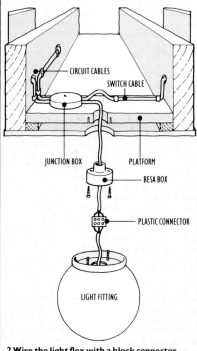

2 Wire the light flex with a block connector

Fittings without backplates

Sometimes close-mounted lights are supplied without backplates.

Wiring Regulations recommend that all unsheathed conductors and terminals must be enclosed in a non-combustible housing, so if you use a fitting with no backplate you must find a means of complying. The best way is to fit a BESA box (1), a plastic or metal box that is fixed into the ceiling void so as to lie flush with the ceiling.

Screw-fixing lugs on the box should line up with the fixing holes in the light fitting's coverplate, but check that they do so before buying the box. You will also need two machine screws of the right thread for attaching the light to the BESA box.

Check that there is no joist right above where you wish to fit the light. If there is one, move the light to one side until it fits between two joists. Hold the box against the ceiling, trace round it and with a padsaw carefully cut the traced shape out of the ceiling.

Cut a fixing board from 25mm (1in) thick timber to fit between the joists and place it directly over the hole in the ceiling while an assistant marks out the position of the hole on the board from below, then drill a cable-feed hole centrally through the marked-out shape of the ceiling aperture on the board. This hole must also be able to take any boss on the back of the BESA box. Position the box and screw it securely to the board.

Have your assistant press some kind of flat panel against the ceiling and over the aperture. Fit the BESA box into the aperture from above so that it rests on the panel, mark the level of the fixing board on both joists, then screw a batten to each joist to support the board at that level.

Fix the board to the battens and feed the cable through the hole in the centre of the BESA box. The light fitting will probably have a plastic connector for attaching the cable conductors (2), and this may have three terminals. Alternatively, a separate terminal for the earth conductor may be attached to the coverplate.

When the conductors are secured fix the coverplate to the BESA box with the machine screws.

If the original ceiling rose was fed by more than one cable, connect them to a junction box in the ceiling void as described above left.

FITTING A DOWNLIGHTER

Decide where you want the light, check from above that it falls between joists, then use the cardboard template supplied with all downlighters to mark the circle for the aperture on the ceiling. Drill a series of 12mm (½in) holes just inside the perimeter of the marked circle to remove most of the waste, then cut it out with a padsaw.

Bring a single lighting circuit cable from a junction box (◁) through the opening and attach it to the downlighter, following the manufacturer's instructions. You may have to fit another junction box into the void to connect the circuit cable to the heat-resistant flex attached to the light fitting.

Fit the downlighter into the opening and secure it there by adjusting the clamps that bear on the upper, hidden surface of the ceiling.

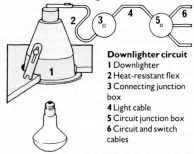

Downlighter circuit
1 Downlighter
2 Heat-resistant flex
3 Connecting junction box
4 Light cable
5 Circuit junction box
6 Circuit and switch cables

FITTING TRACK LIGHTING

Ceiling fixings are supplied with all track lighting systems. Mount the track so that the terminal block housing at one end is situated where the old ceiling rose was fitted. Pass the circuit cable into the fitting and wire it to the cable-connector provided.

If the circuit is a loop-in system mount a junction box in the ceiling void (◁) to connect the cables.

Make sure that the number of lights you intend to use on the track will not overload the lighting circuit, which can supply a maximum of twelve 100W lamps or the equivalent.

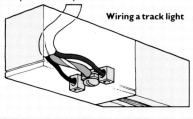

Wiring a track light

Fluorescent lights

Fluorescent light fittings are supplied with terminal blocks for connection to the mains supply.

With the ceiling rose removed screw the fitting to the ceiling, positioned so that the circuit cable can be fed into it conveniently. The terminal block will take only three conductors, so the fitting must be connected to a junction-box system, or a junction box must be installed in the ceiling void to accommodate loop-in wiring as for a close-mounted light (See opposite).

Fluorescent lights normally need earth connections, so they cannot be used on old systems that have not got earth conductors.

You can mount a fluorescent unit by screwing directly into ceiling joists or into boards nailed between joists to provide secure fixings.

Wiring a fluorescent light fitting
A simple plastic block connector is fitted inside a fluorescent light fitting for the circuit cable.

FLUORESCENTS UNDER CUPBOARDS

You can fit fluorescent lighting under kitchen cupboards to illuminate the work surfaces below, the power being supplied from a switched fused connection unit (▷) fitted with a 3amp cartridge fuse.

You can install a second fluorescent light fitting and supply its power by wiring it into the terminal block of the first one.

LIGHT SWITCHES

The commonest type of switch for controlling lighting is the plateswitch. It has a switch mechanism mounted behind a square faceplate that may have one, two or three rockers. Though these are usually quite adequate for domestic use there are also double faceplates with four or six rockers.

A one-way switch simply turns a light on and off, but two-way switches are wired in pairs so that the light can be controlled from two places – typically, the head and foot of a staircase. There is also an intermediate switch that allows a light to be controlled from three places.

Any switch can be flush-mounted in a metal box that is buried in the wall, or surface-mounted in a plastic pattress. Boxes 16 and 25mm (⅝ and 1in) deep are available to accommodate switches of different depths.

Where there is not enough room for a standard switch a narrow architrave switch can be used. There are single ones and double versions that have their rockers one above the other.

A dimmer switch is a device by which the intensity of light can be controlled as well as switching on and off. In some versions a single knob works as both switch and dimmer. Others have a separate one for switching so that the light level does not have to be adjusted each time the light is switched on.

A conventional switch cannot be mounted within reach of a bath or shower unit, and in such situations ceiling-mounted double-pole switches with pull-cords are installed.

Fixing and cable runs

Lighting cable is run underneath floorboards or within the hollow of cavity walls, or is buried in wall plaster (▷). The mounting boxes and switches are fixed to various walls by exactly the same methods as used for sockets (▷).

Light switches must be installed in relatively accessible positions, which normally means at about adult shoulder height for a wall switch and just inside the door of a room.

TYPES OF SWITCHES

Most light switches are made from white plastic but there are some more striking finishes to compliment your decorative scheme. Bright primary-coloured switches can be matched with coloured flex, plugs and socket outlets to make an unusual and attractive feature. Brass antique-reproduction switches suit traditional interiors.

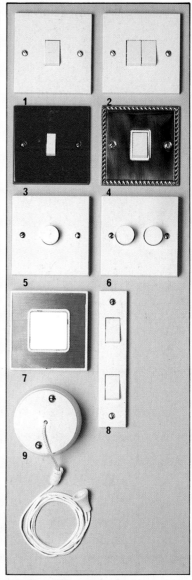

Selection of light switches
1 One-gang rocker
2 Two-gang rocker
3 Switches are made in a range of colours
4 Reproduction antique switch
5 One-gang dimmer
6 Two-gang dimmer
7 Touch dimmer
8 Two-gang architrave switch
9 Ceiling switch

REPLACING SWITCHES

Replacing a damaged switch is a matter of connecting the existing wiring to the new switch in exactly the same way as it was connected to the old one.

Always turn off the power and remove the fuse before you take off the faceplate to inspect the wiring.

In the case of a surface-mounted switch make sure that a new faceplate will fit the existing pattress; otherwise you will have to replace both parts. If you do use the old pattress, also use the old machine screws for fixing on the faceplate. In this way you know that you have matching threads on the screws.

To replace a surface-mounted switch with a flush-mounted one remove the old switch, then hold the metal box over the position of the original switch and trace round it. Cut away the plaster to the depth of the box and screw it to the brickwork (◁). Take great care not to damage the existing wiring while you are working.

Replacing a one-way switch

A one-way switch will be serviced by a two-core and earth cable, and the earth conductor, where there is one, will be connected to an earth terminal on the mounting box. The red and black wires will be connected to the switch itself.

A true one-way switch has only two terminals, one above the other, and the red or black conductors can be connected to either terminal **(1)**. The back of the faceplate will be marked 'top' to ensure that you mount the switch right way up, so that the rocker is depressed when the light is on. The switch would work just as well upside down but the 'up for off' convention is a good one as it tells you that a switch is on or off even when a bulb has failed.

Occasionally you will find a switch that is fed by a two-core and earth cable and operates as a one-way switch, yet has three terminals **(2)**. This is a two-way switch wired up for a one-way function, something that is fairly common and perfectly safe. With the switch mounted right way up the red and black wires should be connected to the 'Common' and 'L2' terminals.

Replacing a two-way switch

A two-way switch will have at least one conductor in each of its three terminals. Without going into the complexities of two-way wiring at this stage, the simplest method for replacing a damaged two-way switch is to write down a note of which wires run to which terminals before disconnecting them. Another way is to detach the conductors from their terminals one at a time and connect each one to the corresponding terminal on the new two-way switch before dealing with the next conductor.

Two-gang switches

A two-gang switch is two single switches mounted on one faceplate. Each switch may be wired differently. One may be working as a one-way switch and the other as a two-way **(3)**.

Use one of the methods described above, for replacing a two-way switch, to transfer the wires from the old to the new terminals, working on one switch at a time.

Replacing a rocker switch with a dimmer switch

Examine the present switch to determine the type of wiring that feeds it and buy a dimmer switch that will accommodate it. The manufacturers of dimmer switches provide instructions with them, but the connections are basically the same as for ordinary rocker switches **(4)**.

HOW SWITCHES ARE WIRED

It is very easy to replace a damaged switch or swap one for a switch of a different nature (◁). The illustrations below show four common methods of wiring switches. If your switch is wired differently it is probably part of a two- or three-way lighting system (◁). Replace the switch as described left.

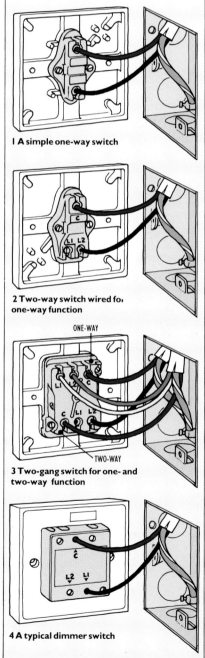

1 A simple one-way switch

2 Two-way switch for one-way function

3 Two-gang switch for one- and two-way function

4 A typical dimmer switch

ADDING NEW SWITCHES AND CIRCUITS

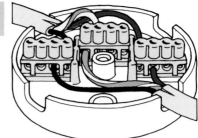

When you want to move a switch or install one where none existed before you will have to modify the circuit cables or run a new spur cable from the existing lighting circuit to take the power to where it is needed.

Replacing a wall switch with a ceiling switch

Light switches must be out of reach of anyone using a bath or shower. If your bathroom has a wall switch that breaks this rule replace it with a ceiling switch that operates by a pull-cord.

Turn the power off at the consumer unit and remove the old switch. If the cable running up the wall is surface-mounted or in a plastic conduit you can pull it up into the ceiling void. It should be long enough to reach the point where the new switch is to be.

If the switch cable is buried in the wall trace it in the ceiling void and cut it, then wire the part that runs to the light into a three-terminal junction box fixed to a joist or to a piece of wood nailed between two joists (▷). Connect the conductors to separate terminals (**1**),

and from those terminals run matching 1mm² two-core and earth cable to the site of the ceiling switch.

Bore a hole in the ceiling to pass the cable through to the switch. Screw the switch to the joist if the hole is close enough; otherwise fix a support board between joists.

Knock out the entry hole in the switch backplate, pass the cable through it and screw the plate to the ceiling.

Strip and prepare the ends of the conductors, connecting the earth to the terminal on the backplate. Connect the red and black conductors to the terminals on the switch – either wire to either terminal (**2**) – then attach the switch to the backplate and make good any damage done to the plasterwork.

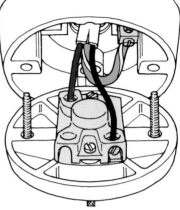

1 Link the switch cable with a junction-box

2 Wiring a ceiling switch

Adding a new switch and light

Turn the power off at the consumer unit and check your lighting circuit to see if it is earthed. If there is no earth wire get expert advice before you try to install a new light.

Decide on where you want the fitting and bore a hole through the ceiling for the cable. Screw a ceiling rose to a nearby joist or nail a board between two joists to provide a strong fixing for the rose (▷).

Bore another hole in the ceiling right above the site of the new switch and as close to the wall as possible. Push twists of paper through both holes so that you can find them easily from above.

Screw the switch mounting box to the wall and cut a chase in the plaster for the cable up to the hole already bored in the ceiling (▷).

Your new light can be supplied from a nearby junction box, from a ceiling rose that is already on the lighting circuit or, if it is more convenient that way, from a new junction box wired into the lighting circuit cable.

From whichever of these sources you choose run a length of 1mm² two-core and earth cable to the new light position, but do not connect the circuit until the whole of the installation is complete. Push the end of the cable through the hole in the ceiling and identify it with tape (**1**). Write 'Mains' on the tape to be absolutely sure. Now

run a similar cable from the switch to the same lighting point (▷).

Strip and prepare the cable at the switch, connecting the earth wire to the terminal on the mounting box, and connect the red and black wires, either wire to either terminal (▷) if it is a one-way switch. If you can get only a two-way switch connect the wires to its 'Common' and 'L2' terminals (▷). It will work just as well. Now screw the switch to the mounting box.

Knock out the entry hole in the ceiling rose, feed both cables through it and screw the rose to the ceiling.

Take the cable marked 'Mains' and connect its red conductor to the central live block, its black one to the neutral block. Slip a green/yellow sleeve over the earth wire and connect it to the earth terminal (▷).

Connect the red wire of the switch cable to the live block and the black wire to the switch-wire block after marking the black wire with red tape. Connect the switch earth wire to the same earth terminal (▷). Screw the cover on the rose.

Make sure that the power is switched off and connect the new light circuit to the old one at the rose or junction box. The new conductors will have to share terminals already connected: red to live, black to neutral and earth to earth (**2**).

Switch on and test the new circuit.

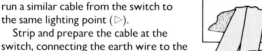

1 Identify the cables

Circuit for a new light ▶
Take the power for a new light from an existing ceiling rose, or insert a new junction box into the existing lighting circuit.

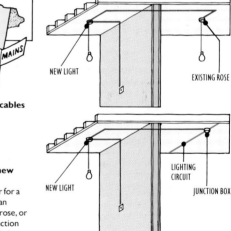

NEW LIGHT

EXISTING ROSE

NEW LIGHT

LIGHTING CIRCUIT

JUNCTION BOX

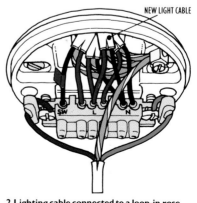

NEW LIGHT CABLE

2 Lighting cable connected to a loop-in rose

ADDING TWO- OR THREE-WAY LIGHTING

Adding a two-way light

There are several situations in which a light should be controllable from two points. A hall light is best switched from both ends of the passageway, and a landing light must be controlled from top and bottom of the stairs.

Installing a new two-way light is very similar to installing a one-way, the only real difference being in the wiring of the switches.

Mount the ceiling rose and both two-way switches, then run 1mm² two-core and earth cable from the power source to the light and from the light to the nearest switch. Don't connect the new installation to the circuit until all the wiring is completed.

Run a 1mm² three-core and earth cable from the first switch to the second, strip and prepare the conductors for connecting to the switches and slip insulating sleeves over the bare ends of the earth wire.

At the first switch you have two cables to connect: the one from the light and the one linking the switches. The light cable has three conductors –

red, black and green/yellow; the linking cable has four – red, yellow, blue and green/yellow. Take the two green/yellow wires, twist their bare ends together and connect them to the earth terminal on the mounting box (**1**).

Connect the red wire from the linking cable to the 'Common' terminal on the switch. Twist together the ends of the yellow wire and either the red or black light cable wire, and connect them to the 'L1' terminal. Twist together the ends of the blue wire and the remaining light cable wire and connect them to the 'L2' terminal (**1**). Screw the switch faceplate to the mounting box.

At the second switch, connect the linking cable's green/yellow wire to the earth terminal, its red wire to the 'Common' terminal, its yellow wire to 'L1' and its blue wire to 'L2' (**1**). Screw the switch's faceplate to the box.

Make sure that the power is switched off and then connect the installation to the lighting circuit at a ceiling rose or a junction box (◁). Test the new installation and repair the plasterwork.

Three-way lighting

You can control a light from three places by adding an intermediate switch to the circuit described above.

The third switch interrupts the three-core and earth cable linking the other two. It has two 'L1' terminals and two 'L2' ones.

At its mounting box you will have two identical sets of wires – red, yellow, blue and green/yellow. Connect the green/yellow wires to the earth

terminal on the box (**2**) and join the two red wires, which play no part in the intermediate switching, with a plastic block-connector (**2**). Ease the block to one side so as to clear the switch when you fit it.

Connect the blue and yellow wires of either cable to the 'L1' terminals on the new switch and those of the other cable to the 'L2' terminals (**2**). Screw the faceplate to the mounting box.

Two-way lighting Circuit (Right)
1 Consumer unit
2 Light fitting
3 Lighting circuit cable
4 Two-way circuit cable
5 Switch
6 Linking cable
7 Junction box

Three-way lighting Circuit (Far right)
1 Consumer unit
2 Light fitting
3 Lighting circuit
4 Three-way circuit cable
5 Switch
6 Intermediate switch
7 Linking cable
8 Junction box

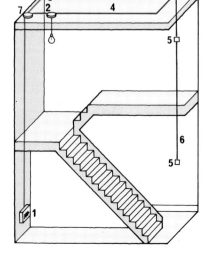

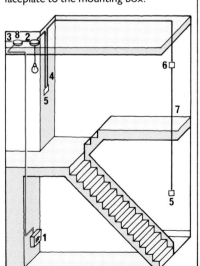

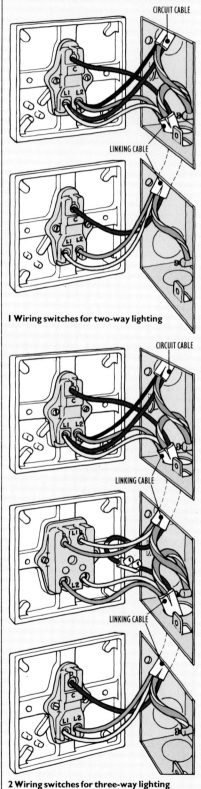

1 Wiring switches for two-way lighting

2 Wiring switches for three-way lighting

ADDING WALL LIGHTS

Many wall lights are supplied without integral backplates to enclose the bared wires and connections. To comply with Wiring Regulations such a fitting must be attached to a non-combustible mounting like a BESA box (▷), a round plastic or metal box which is screwed to the wall in a recess chopped out of the plaster and brickwork.

Alternatively use an architrave switch mounting box. This is a slim mounting that leaves plenty of room on each side for wall plug fixings for the light. Both mountings are fixed to the wall as described for flush socket outlet mountings (▷).

The basic circuit and connections

The simplest way of connecting wall lights to the lighting circuit is via a junction box (▷). The procedure is: complete the wall-light installation, switch off the power, then connect the new installation with the junction box.

Wire up a one-way switch (▷). All the wall lights will be controlled by this switch, though if you choose lights that have their own integral switches they can also be controlled individually.

Run a 1mm² two-core and earth cable from the junction box, looping in and out of each wall-light mounting and on to the last, where the cable ends.

Instead of cutting the cable at the light mountings strip off the outer sheathing, leaving the conductors intact, then carefully slice about 18mm (¾in) of insulation from the middle of each one and pinch the bared wire into a tight bend with pliers. At each light slip green/yellow sleeving over the doubled earth wire and connect it to the earth terminal on the mounting box (1).

Connect the doubled red and black wires to the block-connector inside the wall light, the black one to the terminal holding the blue wire and the red one to the terminal holding the brown wire (1).

The last wall-light mounting will have one end of the cable entering it, so strip and prepare the ends of the wires, then connect them as described above.

1 Wiring a typical wall light

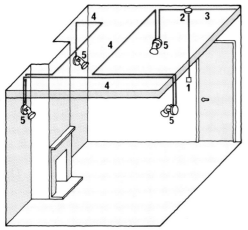

Basic wall-light circuit
The basic circuit and connections are as described above.
1 Switch
2 Junction box
3 Loop-in lighting circuit cable
4 Wall-light circuit cable
5 Wall light

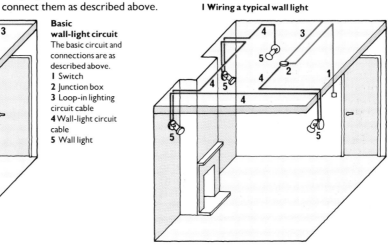

Replacing a ceiling light
You can dispense with a ceiling light in favour of wall lights and use the existing wiring and switch. Switch off the power, remove the rose and connect the wiring to a fixed junction-box (▷).
1 Existing switch and cable
2 Junction box replaces rose
3 Loop-in lighting circuit cable
4 1mm² wall-light cable
5 Wall-light

Ceiling light plus wall lights
If you want to keep your ceiling light you can substitute a double-gang switch for the single one (▷), then wire the present ceiling-light cable to one half of the switch and the new wall-lighting cable to the other.
1 Double-gang switch
2 Old switch cable
3 New switch cable
4 Ceiling light
5 Loop-in circuit
6 Junction box (▷)
7 1mm² lighting cable
8 Wall light

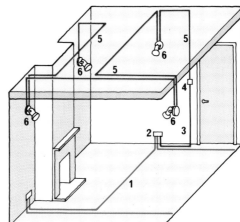

Using a spur
Wall lights can be wired to a ring circuit by means of a spur cable. Run a 2.5mm² two-core and earth spur from a nearby socket to a switched fused connection unit that has a 3amp fuse (▷).
1 Ring circuit
2 Socket outlet
3 Spur cable
4 Fused connection unit
5 1mm² lighting cable
6 Wall light

RUNNING POWER TO OUTBUILDINGS

The power supply to a separate workshop, garage or toolshed cannot be tapped from other domestic circuits.

The cable must run from its own fuseway or switchfuse unit and pass *safely underground or overhead to the outside location, where it is wired into another switchfuse unit from which the various circuits in the outbuilding can then be distributed as required.*

Types of cable permitted outdoors

Three types of cable can be used outside. The one you choose will depend on how you wish to run it.

Armoured
This two- or three-core cable is insulated in the ordinary way but is also protected by a steel-wire armour which is itself insulated with an outer sheath of PVC. In the two-core cable the wire armour provides the path to earth, but some local authorities insist on the third, earth conductor.

Armoured cable is expensive, and must be terminated at a special junction box at each end of its run where it can be connected to ordinary PVC-insulated cable. It is fitted with threaded glands for attaching it to the junction boxes. This type of cable needs no further protection if it is run under the ground.

Mineral-insulated and copper-sheathed
The only other cable that can be buried without further protection is mineral-insulated, copper-sheathed (MICS) cable, whose PVC-insulated conductors are tightly packed in magnesium-oxide powder within a copper sheathing. The copper sheath can act as the earth conductor, and is itself sheathed in PVC insulation.

Because the mineral powder will absorb moisture special seals must be fitted at the ends of the cable when you order it.

MICS cable is expensive and, like armoured cable, must be terminated at junction boxes so that cheaper cable can be used in the outbuilding itself.

PVC-insulated and -sheathed
Ordinary two-core and earth PVC-insulated cable can be run underground to an outbuilding if it is protected against damage by being run through an impact-resistant plastic conduit. For turning corners elbow joints are cemented onto the ends of the straight runs of conduit. The electrical cable itself can be continuous.

This is a much cheaper way of taking power to an outbuilding than by armoured or MICS cable. PVC-insulated cable can also be run overhead quite safely under certain specified conditions (See below).

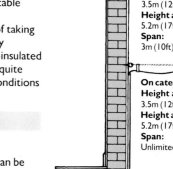

Outdoor cables
1 Armoured cable
2 Mineral-insulated, copper-sheathed cable
3 PVC-insulated and sheathed cable

Ways of running outdoor cable

Underground
Running cable underground is the best way of supplying power to an outbuilding.

You should bury the cable in a trench at least 500mm (1ft 8in) deep – deeper still if the cable has to pass under vegetable plots or other areas where digging is likely to go on. It's best to plan your cable run so as to avoid such areas as much as possible, but you can provide extra protection for the cable by laying housebricks along both sides of it to support a covering made from pieces of paving slab. Line the bottom of the trench with finely sifted soil or sand, lay the cable, then carefully fill in.

Overhead
Ordinary PVC-insulated cable can be run from house to outbuilding as long as it is at least 3.5m (12ft) above the ground or 5.2m (17ft) above a driveway that is accessible to vehicles. The cable may not be used unsupported over a distance of more than 3m (10ft), though the same distance can also be covered by running the cable through a continuous length of rigid steel conduit suspended at least 3m (10ft) above the ground or 5.2m (17ft) above a driveway. The conduit itself must be earthed.

Over greater distances the cable must be supported by a catenary wire stretched taut between the house and outbuilding and to which the cable is clipped or hung from slings. The supporting metal wire must be earthed.

PVC-insulated cable can be run through conduit mounted on a wall.

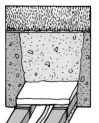

Protecting underground cable
Support paving slabs on bricks to protect a cable at the bottom of a trench.

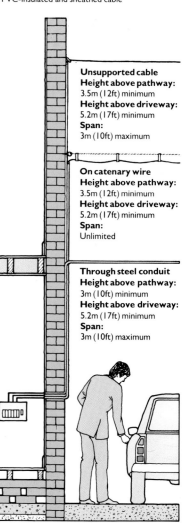

Unsupported cable
Height above pathway:
3.5m (12ft) minimum
Height above driveway:
5.2m (17ft) minimum
Span:
3m (10ft) maximum

On catenary wire
Height above pathway:
3.5m (12ft) minimum
Height above driveway:
5.2m (17ft) minimum
Span:
Unlimited

Through steel conduit
Height above pathway:
3m (10ft) minimum
Height above driveway:
5.2m (17ft) minimum
Span:
3m (10ft) maximum

Running cable overhead

RUNNING THE CIRCUIT

Various equipment and cables can be used to run a circuit to an outbuilding. The method described here uses normal PVC-insulated cable and a switchfuse unit at each end of the circuit, but you could start in a spare fuseway in the consumer unit if one is available.

It is assumed that power sockets and lighting are required in the outbuilding, so the lighting circuit is taken from the power cable via a junction box and an unswitched fused connection unit.

Run the cable underground in impact-resistant plastic conduit, entering both buildings above the DPC but under the floorboards if possible.

House end of the circuit

Install a 30amp switchfuse unit near the meter and fit a 30amp circuit fuse. Mount a residual-current circuit breaker between the unit and the meter.

Run 10mm² two-core and earth cable from the 'Load' terminals of the RCCB to the 'Mains' terminals of the switchfuse unit. Connect the outgoing 4mm² cable to the 'Load' terminals (**1**) of the unit.

Prepare one red and one black 16mm² PVC-sheathed and -insulated cable for the meter leads and attach them to the 'Mains' terminals of the RCCB. Wire a 16mm² earth lead to the RCCB (**1**) ready for connecting to the consumer's earth terminal. Have the Electricity Board make the connections to the meter and earth terminal when the circuit is finished.

Outbuilding end of circuit

Run 4mm² two-core and earth cable through conduit from the house to the outbuilding, terminating at a 30amp switchfuse unit mounted on the wall.

Connect the incoming cable to the supply or 'Mains' terminals of the switchfuse unit and the outgoing 4mm² cable to its 'Load' terminals (**2**), then run this cable to the socket outlets.

(▷) in the outbuilding.

Insert a 30amp junction box at some point along the power cable (**3**) and from it run a spur to an unswitched fused connection unit (▷). Fit a 3amp fuse in the connection unit and run 1mm² two-core and earth cable from the connection unit to the light fitting and switch (▷).

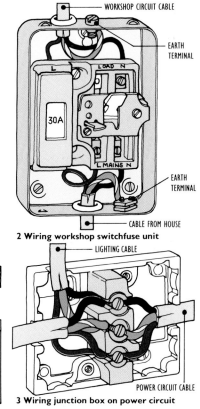

Workshop circuit
1 Meter
2 RCCB
3 Switchfuse unit
4 4mm² cable
5 Conduit
6 Switchfuse unit
7 Junction box
8 Socket outlet
9 Fused connection unit
10 Light fitting
11 Light switch

HOUSE END OF CIRCUIT

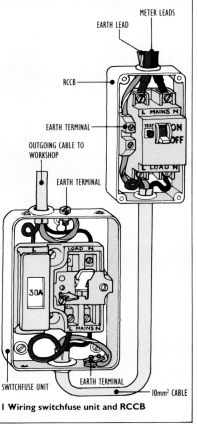

1 Wiring switchfuse unit and RCCB

2 Wiring workshop switchfuse unit

3 Wiring junction box on power circuit

COMPLETE REWIRING

Planning ahead

As an amateur you should carefully consider the time factor before deciding to tackle the complete rewiring of your house. When you are working on only one circuit the rest of the household can function normally, but to install whole new circuitry running to a consumer unit means that eventually every part of the house will be affected.

A full-time professional can cope with all this in such a way that the level of inconvenience to the household is held to a minimum. The amateur, perhaps obliged to work only at weekends, will have to consider a time span of several weeks, especially as it is very important

not to work hastily on such installations. Hurried work can lead to dangerous mistakes.

So unless you are very experienced, and are prepared to make the installation a full-time commitment for a week or two, you would be well advised to employ a fully qualified electrician for this time-consuming job (◁). The expert may be willing to work alongside you, so that you can save considerably on the cost by doing some of the jobs that really have nothing to do with actual electrical work – running cable under floors and channelling out plaster and brickwork, for example.

DESIGNING YOUR SYSTEM

Before meeting your professional you should form clear ideas about the kind of installation you want. Though you may decide between you to change

some of the details a proper specification can help the electrician considerably and make expensive later additions and modifications unnecessary.

CHOOSING THE BEST CONSUMER UNIT

Choose the best consumer unit that you can afford. Cartridge fuses are better than the rewirable ones and are hardly any different in price, while units with miniature circuit breakers are the best of

all, though they are far more costly.

Whichever type you decide to buy, be sure that it has enough spare fuseways for possible additional circuits.

RESIDUAL-CURRENT CIRCUIT BREAKERS

Ask your professional about the value of installing a residual-current circuit

breaker (RCCB) (◁). You could have one built into your consumer unit.

POWER CIRCUITS

For supplying socket outlets ring circuits are better than radial ones. You can have as many sockets as you like providing the floor area in question doesn't exceed 100sq m (120 sq yd), so make sure that your plan includes enough outlets to

meet your present and likely future needs. Trying now to save on the cost of a few sockets can cause you inconvenience in the future, when you may have to start adding spurs.

LIGHTING CIRCUITS

Modern domestic lighting circuits are usually designed round a loop-in system, but remember that for expediency you can supply individual lights from a junction box. You should insist in your plan on a lighting circuit for each floor so

that you will never be totally without electric light if a fuse should blow.

In the interests of safety have two-way or three-way switches installed for the lights in passageways and on landings and staircases.

ADDITIONAL CIRCUITS

If you are having the whole house rewired consider extra radial circuits for

such things as immersion heaters and showers.

CHOOSING STORAGE HEATERS

To be strictly accurate when working out the size and number of storage heaters you'll need you should calculate in the same way as if you were going to install wet-system central heating radiators (see margin, Right). The heat demands of each room are worked out and then heaters are chosen to meet those demands.

Since putting in storage heaters is much easier than installing a full central heating system – a common DIY job, in fact – a guide for the amateur is needed. Such a guide is provided by some manufacturers and the Electricity Boards in the form of simple charts to help you estimate each room's needs on the basis of floor area and the number of outside walls.

Charts like this are not totally accurate, but they allow you to select heaters from the three basic sizes: 1.7, 2.5 and 3.4 kW. A chart of this type is shown below right.

Low-output storage heater
Will provide ample background heat for corridors and hallways.

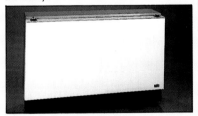

High-output storage heater
Runs on economical night-rate electricity and provides adequate comfort heating for living areas.

Decorative finishes
Many heaters are available with decorative panels and finishes to suit your interior scheme.

STORAGE HEATERS

A sound reason for the popularity of electric storage heaters is their use of cheap off-peak power during the night. The Electricity Boards' special rate for the night-time power is often only half of that charged for the daytime supply, even less. The cut-rate scheme is known as Economy 7, referring to the seven hours of the night when the cheap rate is in force.

The night-time power heats up a core of fire brick or similar material in the storage heater, and the core releases the heat next day, a process of convection in which cool air is drawn in at the bottom of the heater, to be heated as it passes over the hot core and then expelled from the top.

Types of storage heater

The early storage heaters were bulky and space-consuming, and they emitted heat at a set rate so that the user had no control over the output. The heat stored at night could be adjusted, but this required an estimate of the next day's heating needs, and a sudden change of weather could leave the user with too much heat or too little.

Some recent advances in storage heater designs have improved the units. Modern ones are slimmer, some being only 150mm (6in) deep, and they can be wall-mounted or freestanding.

The new heaters have better insulation and allow greater control of the heat output. Adjustable dampers and fans, some thermostatically controlled, allow heaters to be run at low levels in unoccupied rooms and opened up when needed, even late in the day, when older storage heaters run out of stored heat. Some units retain a residue of the stored heat which reduces overnight charging and cuts costs further. Others can monitor room temperatures at night, assess the next day's heating needs (a cold night is normally followed by a cold day) and adjust the heat charge accordingly.

Positioning storage heaters

Like radiators, storage heaters should be placed below single-glazed windows to counter draughts and balance the room temperatures. With double glazing they can go anywhere for convenience or the best heat spread. The heaters have individual circuits, a separate consumer unit and a special off-peak meter to record their power consumption.

SEE ALSO

Details for: ▷
Heater circuits   40-41

● **Designing central heating**
A central heating designer/installer aims at a system that will heat rooms to the temperatures shown here, assuming an outdoor temperature of −1°C (30°F).

ROOM	TEMPERATURE
Living room	21°C (70°F)
Dining room	21°C (70°F)
Kitchen	16°C (60°F)
Hall/landing	18°C (65°F)
Bedroom	16°C (60°F)
Bathroom	23°C (72°F)

SIZING STORAGE HEATERS			
Room sizes		**Storage heaters required**	
Dimensions	Area	Comfort heating	Background heating
2.5×2.75m(8′×9′)	0-7.4m² (0-80ft²)	1×2.5kW	1×2.5kW
3.0×3.4m(10′×11′)	7.4-11.2m²(80-120ft²)	1×3.4kW	1×2.5kW
3.0×4.0m(10′×13′)	11.2-13.0m²(120-140ft²)	1×1.7kW and 1×2.5kW	1×2.5kW
2.75×4.9m(9′×16′)	13.0-14.9m²(140-160ft²)	1×1.7kW and 1×2.5kW	1×3.4kW
4.0×4.0m(13′×13′)	14.9-16.7m² (160-180ft²)	2×2.5kW	1×3.4kW
3.7×4.9m(12′×16′)	16.7-18.6m²(180-200ft²)	2×2.5kW	1×3.4kW
3.7×5.5m(12′×18′)	18.6-20.4m²(200-220ft²)	2×2.5kW	1×1.7kW and 1×2.5kW
3.4×6.4m(11′×21′)	20.4-22.3m²(220-240ft²)	1×2.5kW and 1×3.4kW	1×1.7kW and 1×2.5kW
4.9×4.9m(16′×16′)	22.3-24.2m²(240-260ft²)	1×2.5kW and 1×3.4kW	1×1.7kW and 1×2.5kW
4.6×5.5m(15′×18′)	24.2-27.9m²(260-300ft²)	2×3.4kW	2×2.5kW
4.3×7.4m(14′×24′)	27.9-31.6m²(300-340ft²)	3×2.5kW	2×2.5kW
5.2×6.4m(17′×21′)	31.6-33.4m²(340-360ft²)	3×2.5kW	2×2.5kW
4.9×7.6m(16′ x 25′)	33.4-38.1m²(360-410ft²)	2×2.5kW and 1×3.4kW	1×2.5kW and 1×3.4kW

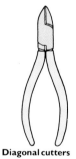

Torch
Keep a torch handy for checking your consumer unit when a fuse blows on a lighting circuit. You may also need artificial light when working on connections below floorboards or in a loft, and a torch that stands unsupported is very helpful.

Diagonal cutters

● **Essential tools**
Terminal screwdrivers
Wire cutters
Wire strippers
Power drill and bits
Electronic mains tester
Torch
General-purpose tools

ELECTRICIAN'S TOOL KIT
You need only a fairly limited range of tools to make electrical connections. The largest number is needed for making cable runs and fixing accessories and appliances to the house structure.

SCREWDRIVERS

Buy good screwdrivers for tightening electrical terminals. The cheap ones are practically useless, being made from such soft metal that the tips soon twist out of shape.

Terminal screwdriver
A terminal screwdriver has a long, slim cylindrical shaft ground to a flat tip. For work on the terminals in sockets and larger appliances buy one with a plastic handle and a plastic insulating sleeve on its shaft. Use a smaller driver with a very slim shaft to work on ceiling roses or to tighten plastic terminal blocks on small fittings.

Cabinet screwdriver
You will need a woodworking screwdriver to fix mounting boxes to walls.

SPANNER

A small spanner is needed for making the earth connections in some appliances and for supplementary earth bonding (◁).

WIRE CUTTERS

Use wire cutters for cropping cable and flex to length.

Electrician's pliers
These are engineer's pliers with insulating sleeves shrunk onto their handles. You'll need them to cut circuit conductors and to twist their ends together.

Diagonal cutters
Diagonal cutters will crop thick conductors more effectively than electrician's pliers. To cut meter leads you may need a junior hacksaw.

WIRE STRIPPERS

There are various tools for removing parts of the plastic insulation that covers cables and flexible cords.

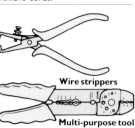

Wire strippers

Multi-purpose tool

Wire strippers
To remove the colour-coded insulation from cable and flex, use a pair of wire strippers with jaws shaped to cut through the plastic without damaging the wire core. There is a multi-purpose version that can both strip the insulation and crop the wires to length.

Sharp knife
Use a knife with sharp disposable blades for slitting and peeling the outer sheathing on conductors.

DRILLS

When you run circuit wiring you need a drill and several special-purpose bits for boring through wood and masonry.

Joist brace
A joist brace has a chuck that takes standard brace bits but its side-mounted handle and ratchet mechanism allow drilling in the restricted space between floor and ceiling joists.

Auger
Some electricians use a long wood-boring auger to drill through the wall head plate and noggings when running a switch cable from an attic down to its mounting box, but it's hard to use such a long tool in the restricted roof space of a small modern house or apartment.

Power drill
A power drill is best for boring cable holes through timbers and making wall plug fixings for mounting boxes. As well as a standard masonry bit for wall fixings you will need a much longer version for boring through brick walls and clearing access channels behind skirting boards.

If you shorten the shaft of a wide-tipped spade bit you can use it in a power drill between floor joists instead of hiring a special joist brace.

TESTERS

Even when you have turned off the power at the consumer unit use a tester to check that the circuit is 'dead'.

Electronic mains tester
An electronic mains tester has a light in its handle that glows when its screwdriver-like tip touches a 'live' wire or terminal. You have to place a fingertip on the tool's metal cap for the light to work but there is no danger of a shock. A small test button on the handle tells if the tool is in working order.

Continuity tester
A continuity tester will test whether a circuit is complete or an appliance is properly earthed. You can buy a tester or make one by connecting a 9-volt battery and a bulb via short lengths of flex and crocodile clips.

Using a continuity tester
Switch off the power at the consumer unit before making the test. To find the two ends of a buried disconnected cable, twist the black and red conductors together at one end, then attach the clips to the same conductors at the other end (1). The bulb should light up. Untwist conductors. If the bulb goes out, the two ends belong to the same cable.

To check that a plug-in appliance is safely earthed, attach one clip to the earth pin of the plug—the longest of the three—and touch the metal casing of the appliance with the other clip (2). If the earth connection is good the bulb will glow brightly; a dimly

glowing bulb indicates poor earthing, which should be checked professionally.

Don't use the appliance if the bulb lights up when you attach the clip to either of the other pins (3). It is dangerous and should be overhauled professionally. Make sure the plug fuse is working.

You cannot test a double-insulated appliance (◁) as it has no earth connection in the plug.

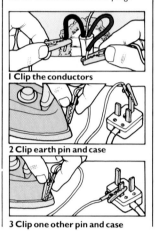

I Clip the conductors

2 Clip earth pin and case

3 Clip one other pin and case

GENERAL-PURPOSE TOOLS

Every electrician needs tools for lifting and cutting floorboards, for fixing mounting boxes and for cutting cable runs through a building.

Claw hammer
For nailing cable clips to walls and timbers.
Club hammer
For use with a cold chisel.
Cold chisel
For cutting channels in plaster and brickwork so as to bury cable or mounting boxes.
Bolster chisel
For levering up floorboards.
Wood chisels
For notching floor joists.
Padsaw or power jigsaw
For cutting through floorboards close to skirtings.
Floorboard saw
A floorboard saw is best for cutting across a prised-up board, though a tenon saw is a reasonable substitute.
Spirit level
For checking that accessory mounting boxes are being fixed horizontally.
Plasterer's trowel or filling knife
Use either tool for covering concealed cable with plaster or other filler.

GLOSSARY OF TERMS

Accessory
An electrical component permanently connected to a circuit – a switch, socket outlet, connection unit etc.

Adaptor
A device used to connect more than one appliance to a socket outlet.

Ampere (Amp)
A unit of measurement of the flow of electric current necessary to produce the required wattage for an appliance.

Appliance
A machine or device powered by electricity.

Catenary wire
A length of wire cable suspended horizontally between two points.

Ceiling rose
A special junction box for connecting a suspended light fitting to a lighting circuit.

Ceiling switch
A light switch attached to a ceiling and operated by a pull-cord.

Chase
A groove cut in masonry or plaster to accept an electrical cable. To cut such grooves.

Circuit
A complete path through which an electric current can flow.

Circuit breaker
Special switch installed in a consumer unit to protect an individual circuit. When a fault occurs, the circuit breaker switches off automatically.

Conductor
A component, usually a length of wire, along which an electric current will flow.

Consumer unit
A box situated near the meter which contains the fuses or MCBs protecting all the circuits. It also houses the main isolating switch which cuts the power to the whole building.

Corrugated plastic sheet
Lightweight PVC sheets used to roof outbuildings and lean-to extensions.

Dimmer switch
A switch which varies the current through a lamp to change the level of illumination.

Double-pole switch
A switch which breaks both the live and neutral conductors.

Downlighter
A type of ceiling-mounted light fitting which directs a relatively narrow beam of light to the floor.

Draw wire
A flexible wire or steel tape used to pull electric cable through narrow and confined spaces.

Earth
A connection between an electrical circuit and the earth (ground). A terminal to which the connection is made.

ELCB
Earth-leakage circuit breaker. See RCCB – Residual-current circuit breaker.

Extension
A length of electrical flex for temporarily connecting the short permanent flex of an appliance to a wall socket.

Fuse
A protective device containing a thin wire which is designed to melt at a given temperature caused by an excess flow of current on a circuit.

Fuse board
Where the main electrical service cable is connected to the house circuitry.
The accumulation of consumer unit, meter etc.

Grommet
A ring of plastic or rubber lining a hole to protect an electrical cable from chafing.

Immersion heater
An electrical element designed to heat water in a storage cylinder.

Insulation – electrical
Nonconductive material surrounding electrical wires or connections to prevent the passage of electricity.

Insulation – thermal
Materials used to reduce the transmission of heat. Insulating a hot-water cylinder and the roof are the most cost-effective measures.

MCB
Miniature circuit breaker. See Circuit breaker.

Neutral
The section of an electrical circuit which carries the flow of current back to source.
A terminal to which the connection is made.

Nogging
A short horizontal wooden member between studs in a timber-framed wall.

Phase
The part of an electrical circuit which carries the flow of current to an appliance or accessory. Also known as live.

PME – Protective multiple earth
A system of electrical wiring in which the neutral part of the circuit is used to take earth-leakage current to earth.

Radial circuit
A power circuit feeding a number of socket outlets or fused connection units and terminating at the last accessory.

RCCB – Residual-current circuit breaker
A device which monitors the flow of electrical current through the live and neutral wires of a circuit. When it detects an imbalance caused by earth leakage, it cuts off the supply of electricity as a safety precaution.

Ring circuit
A continuous power circuit starting at and returning to the consumer unit.
Also known as ring main.

Rising main
The pipe which supplies water under mains pressure, usually to a storage cistern in the roof.

Rocker switch
A modern-style switch operated by a lever which pivots about its centre.

Sheathing
The outer layer of insulation surrounding an electrical cable or flex.

Short circuit
The accidental rerouting of electricity to earth which increases the flow of current and blows a fuse.

Spotlight
A light fitting which directs a narrow beam of illumination onto a specific object or area.

Spur
A short length of cable feeding a socket outlet or fused connection unit and taking its power from another similar accessory.

Storage heater
A space-heating device which stores heat generated by cheap night-rate electricity, then releases it during the following day.

Studs
The vertical members of a timber-framed wall.

Supplementary bonding
The connecting to earth of exposed metal appliances and pipework in a bathroom or kitchen.

Terminal
A connection for an electrical conductor.

Toggle switch
An old-fashioned switch operated by a projecting lever.

Transformer
A device which increases or decreases voltage on a circuit.

Uplighter
A light fitting which reflects illumination onto a ceiling.

Volt
A unit of measurement of 'pressure' provided by Electricity Board generators that drives the current along the conductors.

Wall fixings
Fibre or plastic plugs providing screw-fixing points for holding mounting boxes or light fittings to a solid masonry wall.
Hollow-wall fixings provide secure fixing points on a timber-framed wall by expanding in the void behind a plasterboard cladding.

Wiring Regulations
A code of professional practice laid down by the Institution of Electrical Engineers.

Page numbers in *italics* refer to photographs and illustrations

INDEX